THE NARRATIVE BRAIN

FRITZ BREITHAUPT

The
Narrative
Brain

The Stories Our Neurons Tell

Yale UNIVERSITY PRESS

NEW HAVEN AND LONDON

Translated and revised by the author.

Yale University Press books may be purchased in quantity for educational, business, or promotional use. For information, please e-mail sales.press@yale.edu (U.S. office) or sales@yaleup.co.uk (U.K. office).

Set in Janson type by Integrated Publishing Solutions. Printed in the United States of America.

ISBN 978-0-300-27380-9 (hardcover : alk. paper) Library of Congress Control Number: 2024941358 A catalogue record for this book is available from the British Library.

This paper meets the requirements of ANSI/NISO Z39.48-1992 (Permanence of Paper).

10 9 8 7 6 5 4 3 2 1

The book is dedicated to the memory of my parents:
Friedrich Carl Breithaupt (1935–1977) and
Ilse Breithaupt, née Hoffmann (1935–2021)

CONTENTS

THE NARRATIVE BRAIN

INTRODUCTION

As humans, we have overcome an incredible obstacle. We can share our experiences with others. Consequently, we are not alone. We share the world. This book is about the very human ability of sharing experiences.

To be sure, animals also coordinate group behavior and utilize collective group intelligence in many ways. A wide range of non-human animals can communicate emotions and share relevant information with others. When one animal has undergone a life-changing experience, others can perceive the change and react to the individual's emotional states. However, even then the other animals do not know specific details about what happened. The individual experience remains hidden and resists sharing.

This is not the case for humans. We have solved the problem of sharing experiences and are able to communicate what has happened to us. This experience of sharing goes beyond simply stating facts and exchanging information. We can relate the past event in such a way that others can co-experience the event step by step and even be changed by that co-experience.

This is an incredible ability. We are not locked into our own minds. Your experience can become my experience. Your sadness can become mine, and this can help both of us. Co-experience is the very opposite of loneliness.

So how do we do this? By means of storytelling and narrative thinking. Stories are everywhere, and that is no surprise given what they allow us to do. By this I mean not only the movies we watch and the books we read but also the many conversations we have about who did what to whom, our posts on social media, our memories, and our own thoughts about what we would do in certain situations. These can seem like little episodes we watch in our minds. We also have our fantasies, the way we remember things, and how we plan events.

In narratives, we relive the experiences of others and share them.

We can also relive our own past experiences or slip into fantastical worlds of things that have never happened. This is possible because in narratives we can imagine ourselves in the place of others and have "their" experiences ourselves. We don't have to touch a hot stove, rob a bank, or cheat on our partner to realize that maybe that's not such a great idea. Something in us holds us back from doing these things, and it's not morality or knowing better, but somehow already having had a related experience that was given to us in narrative form. Narratives give us the pleasure of trying out the forbidden.

There is a common saying, "You can't have your cake and eat it too." But with narratives we can do just that: we can have experience—narratively, mentally—and at the same time take no part in the action. Yes, there are many things I would love to imagine but will not try out in real life. I assume that the same is true for my readers. By living vicariously, we double our lives. We co-experience what has happened to others. Stories allow us to witness what has already happened in the past, which we can thereby experience as if it were the present. We can even project things that have not yet occurred and make them available to us.

In this respect, narrative thinking is a great medium for experiencing and planning. You don't have to be an evolutionary biologist to recognize that narrative thinking offers real survival advantages. Narrative thinking better prepares us for future life situations. We learn from each other, and we manage to do something incredible: the experiences of one person can become the experiences of another. We are not singular isolated beings, but a network of individuals. Sponges, ants, and herds of mammals share only a limited range of experiences, such as blind panic in cases of immediate danger. We, on the other hand, are constantly multiplying our experiences and can therefore make better decisions and potentially prepare for a range of events. This allows fewer events to traumatize us.

However, this raises a second question: why do we engage in this narrative thinking? We don't do something just because it gives us selective advantages. After all, we don't get involved in the laborious exercise of reproduction with all the problems of finding a partner because it helps our genes to spread, but because sex magically attracts us, and love makes us happy. Conversely, not everything we do makes sense in evolutionary terms. For example, I doubt that the foods most popular in the United States are good for us.

This second question—why we engage in narrative thinking—is the starting point for this book. What's in it for us? We will come to an answer by examining telephone games, that is, chains in which people retell stories and events, to see what sticks in narratives. In fact, my lab conducted what is likely the largest retelling study to date, with close to 20,000 retellings and 12,800 participants. It turns out that emotions play a central role in this process. Narrative thinking may be so attractive to us because it rewards us with specific emotions. When we co-experience a narrative episode, we also co-experience the emotions of the sequence of events. Having emotions is certainly better than not having emotions, and there is a long list of highly attractive emotions. This is true even for some negative emotions. Another benefit of these emotions is that they have a stop function that allows us to step out of the narrative and return to our present situation. In short, emotions reward us but also release us from the narrative. It seems that we want to dive deep into our dream spaces but also be able to return back to the surface.

This thesis regarding the emotional rewards involved in narratives will lead us to a third question. Narratives are, in a sense, addictive. Or, to put it more cautiously, certain narrative sequences imprint themselves on us to such an extent that we revisit them again and again and get used to them. We all have our own weaknesses in this regard. Some want to see themselves as triumphant heroes, while others conceive of themselves as victims. I'll bet avid readers will know what narrative emotions attract them and become their go-to choices in their daydreams, movie selections, and memories of their lives. This "sticking" tendency of narrative emotions raises the question of whether narrative thinking, which takes us out of our narrow existence and allows us to experience the lives of others, does not also hold us captive. In other words, can we "change our narratives," as people like to say these days? Or are we stuck in a constricted world of narrative patterns that we repeat over and over? Could it be that this ability that lifts us beyond our individual limited lives also backfires and imprisons us?

Here we get to the large question of who we are because we think, co-experience, and live in narratives. We are narrative creatures, but what does this mean?

I Think I Am on the Wrong Track

Everyone is the architect of his or her own life, as the saying goes. This is especially true when it comes to how we perceive ourselves and our place in the world. With a positive outlook, we find happiness everywhere and can be resilient. But this can also go wrong. We all know pessimists for whom even the best news somehow becomes evidence of their unhappiness. We'd often like to grab hold of such a pessimist and shake him to wake up from the bad dream, but it isn't likely to change anything. On the contrary, it would only reinforce their belief that everyone is against them, including the friend shaking them.

Our thinking directs us to the grooves of sameness and repetition. We expect what we know. Obviously, it is not easy to change one's patterns. Like the pessimist, we can all be caught up in our vision of how things are. This determines who we are, where we are, and how we envision our future. Most of all, it means we all keep getting trapped in our narratives. We expect certain things and are entangled in our bleak expectations until they occur. And when they don't occur, we wait until they do. In the process of waiting, we reshape actual events in our minds to fit our vision. In the United States, for instance, after years of discord and contending narratives, many people can't help but view Donald Trump as either a hero or an utterly evil villain. This overall image shapes how people frame every single action or statement that he makes.

As a professor, I know many colleagues who cling to a vision from their teenage years. They want to become a professor at Harvard, win a Nobel Prize, or find a cure for cancer. These are all worthy goals that may spur us on to work. But these visions do not seem to make my colleagues happy. Instead, some seem bitter, frustrated, and envious of what others have accomplished. The narrative these colleagues have chosen no longer fits their lives. But they don't seem to be able to drop it.

Some people constantly see themselves in the role of victim. Of course, it is important to recognize when one is being oppressed, in order to rebel against it or seek help. But being a victim can also become a role that is revisited again and again because someone knows that part too well and can slip into it like a glove that fits and feels right. After all, the narrative portrays victims as morally superior and diminishes their responsibility and agency. For the victim, the narrative can thus be a relief, but one that binds them to more victim narratives. We shall talk more about victim and vulnerability narratives later.

Another variation I often hear from my friends in Germany is the narrative of the one hardworking worker bee and the hundred lazy parasites. Unfortunately, it is not a fairy tale with a good outcome, because my friends find themselves surrounded by exploiters and freeloaders. In the end, the wrong person always wins, even though each of my friends alone keeps the business going as the frustrated worker bee. Listening to this, I always wonder whether these parasites actually exist, since everyone in Germany tells the same tale of being the worker bee.

Likewise, certain family roles can capture us. One of them is the super-mother, a figure found in many cultures. Always smiling, she keeps the household running, becomes the best friend of her children's friends, takes on the emotional labor in the family, manages to always have food ready for everyone in a clean kitchen, and at the same time still maintains a career without complaints. The pressure is great and there is hardly a free minute, especially when something doesn't work out.

Love is another domain full of narrative expectations. You could fall in love with the wrong person. Or you could fall in love with the right person, but the right person turns out to be the wrong person if your love is not reciprocated. Or the beloved turns out to be a narcissist, and the right person is thus the worst choice. However, it is impossible to forget the one you love. Again and again, small sequences flash before you: how you are together in an Argentine tango bar, how you grasp each other's hand, or how it feels to share a joyful moment. Letting go is hard. Unfortunately, some of us seem to fall for similar people, even when it's the wrong person, again and again.

These examples show a wide range of behaviors in which a self-image becomes a trap. At first glance, holding on to such an imaginary self-image is not a matter of narrative thinking. We might try to explain these fixations in terms of worldviews, past experiences, imprints, schemas, scripts, patterns, trauma, or ideals. This may all be true, but at the same time these self-images can only exist because they stand before us as concrete micronarratives with the potential for action. We see ourselves as heroes, as victims, as persuaders, as super-mothers, or as lovers only because we can imagine ourselves in a situation that involves actions and events with a narrative before-and-after structure. Some narratives may stand before us as a guiding path of what will happen in what order, while others just appear to us in a flash of split seconds. Even these short versions offer some development, some potential for action. It is in this way that self-images are tied to unfolding narratives.

It is not easy to leave behind these narratives and the self-images built on them. Narratives are the form in which our brain simulates our actions and the actions of others. Because we consider these simulations suitable to represent our actions, they are suspiciously close to reality and we take them as real. And who wants to say goodbye to reality? But there is also a way out.

On one hand, narratives are representations or simulations of the real world with actions and feelings. On the other, they are merely figments of our imagination with their own rules and forms that do not simply dance to the tune of reality. Narratives have forms with which we categorize the observed actions of ourselves and others. When we observe others, we quickly attribute certain motivations and interests to them; we pin them down. We observe events in small sequences and episodes in which everyone takes on dyadic or triadic roles in relation to each other: villain, perpetrator, hero, love interest, rival, helper, liar, victim, judge, friend, false friend, traitor, sociopath, witness, mentor, freeloader—the list goes on and on, but it is not infinite. These roles only exist in our minds, because of course all people have different sides and might imagine themselves in different roles. However, when we observe a social situation, the fact that we can commit people to one role or another is a tremendously convenient simplification. This makes the situation manageable and opens its potential for simulation. These roles offer a simple code for narratives to unfold. We cast someone in the role of victim, and then we find a perpetrator to blame. Now the story can unfold in our mind. The narrative can run smoothly in our minds, like a movie, and can also be clearly remembered afterward. At that point, we cannot distinguish between reality and story any longer since the story covers anything we have perceived in the real world.

In short, narratives offer a highly attractive orientation in a complex world. Who could say no to that? In any case, our brain does not miss this opportunity. Narratives allow us to make assumptions and predictions about the social world, to remember them, and to share them with others. This is not just convenient but quite rational and usually exceptionally efficient. But it is also precisely how we get stuck in a rut. The narrative, once developed, is so convincing that we cannot simply shake it off.

In many cases, narrative thinking offers quick orientations and good predictions. We know what to expect in certain situations and know what to do. Still, there are also many forms of misbehavior that can be guided by narrative thinking. A short trigger can start an entire chain of behavior.

In a frightening racist incident in my American hometown, on July 4, 2020, a man was confronted by a group of white people who were, allegedly, shouting "Get the noose!" In a chilling way, this phrase put before everyone's mental eyes what sequence of events might have taken place. The words trigger a chain of expectations, not only a narrative pattern but an action program. Such incited narratives often materialize. This possibility was also known to the potential victim, a black man, who for his part might have provoked this situation. (It turns out he had intruded on a party on the group's private property twice in a row, returning a second time after he was escorted out once.)

Although his friends caught the action on their cell phones, the horrific call for a noose was not recorded. Nevertheless, what seems to have started as an accidental trespassing quickly solidified into a narrative program, following century-old narratives of oppression.

To be sure, not all forms of behavior guided by narrative expectations are terrible. A typical behavior in committees or teams, which I know from the university but which is certainly part of everyday life elsewhere, is that someone decides quickly—all too quickly—to be for or against an initiative. After the initial decision, the committee member almost always sticks with it. Whoever is for it can always find justifications, even if rational objections arise. Those who are against can make everything look bad. This is not just about the intellectual convenience of maintaining a judgment once it has been made. Rather, it is essentially about one's role, about how one appears to the committee and how one casts oneself in the group: as an enthusiastic advocate or as a sharp rationalist who seeks out the flaws in the thinking of others. People can cultivate their roles and pepper them with the right formulations, facial expressions, and codes of collegial behavior. Many good ideas and initiatives are destroyed in this way.

The sunk cost fallacy (or escalating commitment) also belongs to the circle of narrative patterns: we have decided on something and now simply cannot let go. We have this concert ticket and are now driving there through a snowstorm, even though we are sick. The alternative of simply staying in bed doesn't occur to us, because the vision of our enthusiastic experience of the concert is too clearly in sight of our mental eyes, even though, in reality, we will hardly be able to enjoy the music with our cold. Perhaps we imagine how we will tell others about our great experience, no matter what it really has been.

Here we come back to the pessimist, the unhappy lover, and the frustrated colleague all holding on to the wrong narrative. What can help them? The answer: more narrative thinking!

Narratives are not one-way streets. Rather, narratives always come in the plural. Every narrative has a counter-narrative. And more important, when we find ourselves in a narrative, whether watching a movie or contemplating the narrative of our lives, things can always turn out differently. Narratives are, as I call it, "multiversional." I am not referring here to several versions of a medieval manuscript. Rather, *multiversionality* means that as we find ourselves entangled in a narrative, we are faced with more than one possible development and outcome. Narratives are so exciting and intense for us because everything could always turn out differently. And we ourselves constantly create this plurality by anticipating what could or should happen. That is, while we think narratively, we are always in the middle, not yet at the end.

It is precisely this mental multiversionality (to throw a beautiful conceptual monster into the field of inquiry) that offers a way out of the ruts that we have dug for ourselves with our narratives. Narratives always allow us to skip and jump onto other tracks simply by imagining doing so. Narratives can be the medium of our unhappiness, but they are also the means of escaping it. The spear that strikes the wound also heals it.

To avoid misunderstandings: this is not a self-help book. I am quite unsuitable as a therapist. If someone were to tell me a personal story, looking for a way out of a misfortune, I, armed with the findings of this book, might perhaps fail to select the best way out of my friend's mess and instead recommend the most exciting storyline. Suspense can be great, but one might not want to have to endure everything that is suspenseful in one's own life.

However, in narrative thinking, everything could always turn out differently—even if it usually does not. Couldn't I have said something else that one day in that little café? We could still be together. Couldn't Ned Stark in *Game of Thrones* have sensed the trap? Couldn't I have convinced my mother to go to the doctor sooner? Why doesn't Elizabeth realize Wickham is a fraud?

Every narrative generates ideas or hopes of alternatives. We imagine how things could have gone differently. We wish our favorite character would suddenly fall out of character to escape a trap. We realize that friends are ensnared in bad narratives. Sometimes we ask ourselves if we should do something crazy right now instead of continuing with the mundane. And at times, we do. Because we think narratively, we can also recognize and at times seize upon *alternative versions*. Narratives allow us to see ways out, even where there seem to be none. We can shape the narratives that play out around us; we can get creative and exercise control. Imagine: we can do what we really want to do. It sounds simple. But to get there, we need narratives that show us where we are and no longer want to be.

This book offers a suggestion on how our narrative brain works. The path of this book combines everyday experiences with works of fiction, and it is also guided by the findings of numerous scientific disciplines. At more than one point, we will dig a little deeper and report on the studies conducted in my laboratory, the Experimental Humanities Laboratory at Indiana University. As the empirical studies conducted there show, stories are not always about clarity and the completion of thoughts but also about confusions, coincidences, and alternatives.

Central Ideas

Life is experienced more intensely when we are enmeshed in stories— I narrate, therefore I am. But it is not only our own lives that are heightened by narratives; through narratives we are also able to transform individual experience into shared experience. In narratives, we reengage with our own past, but we also relive the experiences of others and share them.

Narratives allow us to co-experience the situations of others. This book explores some of the ways in which narratives codify our experiences, thereby bringing order to the chaos of our individual perceptions, contexts, and events. Narrative order helps us to remember and communicate these experiences in such a way that other people can— and want to—experience them, and it also allows us to relive them. Furthermore, as an expressive behavior, our storytelling ability gives us the creative potential to invent experiences, enable narrative empathy, plan future actions, and produce fiction.

Thanks to our narrative brain, we are connected to other beings in a deep way, because we can "undergo" similar experiences just by talking to them. We are not alone in our lives and our most important experiences, since we can share them with others. This is an amazing capacity. Narrative co-experience allows for a community that goes far beyond simply being together at the same time and space. We can escape from the prison of our own brains and the here and now. This ability is connected with something that I call "mobility of consciousness," a mental skill that is a great evolutionary achievement of our species. To understand how our consciousness became mobile, we need to look at narratives, because they offer transportation and immersion into strange worlds.

The list of questions about narrative thinking and the tasks we face is long, but these can basically be divided into three groups:

1. What are narratives—specifically, what are narratives in our thinking? What do they convey? How does the content of narratives differ from other types of information? When we hear or read narratives, what sticks with us? What gives them stability?
2. Why is it attractive for us to engage in narratives? Why is it that we not only engage in this complex way of thinking but have elevated it to one of our most popular leisure activities? What's in it for us?
3. How do narratives propel our consciousness out of the here and now and lift it into the worlds of others? How does the immersion into another world succeed? Where do the intensities of the target world come from that make us (almost) forget our body in our home world? What is mobility of consciousness, anyway?

Let me elaborate on one central idea in advance. Episodes structure narrative thinking. At the end of each successful episode comes an emotion. Narratives are emotion episodes. Emotions reward narrative thinking. Narratives tempt us, detach our consciousness from the here and now, and transport us to a world where we hope for an emotional reward. We also seek this episodic structure with emotional reward in our behavior. Narrative thinking is so important for us because it allows us to recognize the beginning and end of sequences and, with the emotion at the end, gives us a signal that something has been accomplished and completed. The emotion at the end is therefore a reward in a double sense. It rewards and evaluates the concrete actions we have narratively experienced and at the same time rewards us for having engaged in a narrative at all.

Narratives train us to expect emotions. We expect or hope for certain emotions as a reward for our engagement in narrative sequences. The happy ending obviously rewards morally good behavior. We may feel the same way in real life: after hard work we feel we deserve to rest happily on our accomplishments. Adventure and hero stories train us to expect triumph. In love stories, after appropriate delays, we expect erotic fulfillment. Subtle stories accustom us to aesthetic emotions of sudden clarity. The narrative of curiosity finds its reward in the feeling of wonder. But more complex and negative feelings also belong here: the ambivalent feeling of shame can be a punishment for the one who stepped out of line. Shame thus rewards the angry observer. Sometimes, sadness is rewarding too. But here it quickly gets complicated, because we always need to ask from which perspective the scene is witnessed.

These expectations of emotional rewards structure our lives and thus subject us to the narrative thinking in episodes. They can make us addicted. We may be aware of this in the extreme cases from the notorious gambler to the excessive gamer, and the workaholic to the sex addict, but other people also orient their lives toward some doses of narrative emotion.

Basic Terms

I would like to briefly introduce the most important concepts for this book. These basic concepts will be like the pieces of a mosaic from which the overall picture can emerge.

Mobility of consciousness. Our minds possess a miraculous mobility of consciousness, an ability to perceive or imagine experiences or states other than the physical reality being occupied at a given time, including those of others. Put more simply, we can move our mind to situations that are different from the now and here. Narrative thinking is its medium. With narrative thinking, we can mentally live in the situation of any other being and thus co-experience it. By understanding the situations of others as narration, they become accessible for us to experience.

Simulation. With narrations, we simulate actual or fictitious situations. We imagine and even start to feel a body we do not have. On one hand, simulation is an embodied duplication of bodily life, and in some respects, simulations are oriented toward regular processes of life. On the other hand, simulations are disconnected from the world of bodies and the regime of strict temporal succession. They only play out in our mind. This disconnect begins with the fact that we can revisit past events and mentally replay them. We can also plan and fantasize. We can imagine what it would be like to leave everyday life and move to a tropical island. Will I really be happy then, next month, ten years from now? What would dating a movie star actually be like? Some simulation is also much more mundane. In the morning before I get up, I go through my calendar, simulate where I have to be and when, who I will meet there, and what will happen at these appointments. This is usually a pretty boring story. Still, I do take some time to think about the more difficult meetings and at times make them more complex when I rehearse them. In some cases, I play out entire scenes and dialogues from many perspectives.

Such simulations are said to take place "offline," or in a state decoupled from actual and physical action. Offline behavior includes the observation of others but also mental simulation of our own behavior. Neuroscientists are analyzing how the neural networks established "online" are simultaneously used for reenacting and simulating social interactions (re-use).[1] Conversely, it is possible that networks established through simulation or observation anticipate one's own behavior. Such simulations through narrative thinking likely play an important role in many decisions we have to make. They might include, for example, moral decisions, such as wondering if we can get away with a little cheating or how we should respond to moral transgressions by others.

Situation. Narratives place characters into situations. When an army operative states, as in *Avatar* (2009), "we have a situation," this spells out trouble. Noticing that "we have a situation" is a core feature of any narrative. In narratives, we are constantly in a specific situation. We are in the middle, in medias res, not yet able to survey the entire sequence, because not everything has been decided yet. And this is then already the first characteristic of situations in general: they are undecided. They stimulate pressure to make a decision or contain the potential to be decisive. Every situation includes and requires a potential for change.

Situations are also concrete and have a certain clarity with a limited or focused repertoire of possibilities for development (this is true even if we do not know or understand many of the possibilities). In summary, a situation is the state of being of a person or character in a transitory space that will change and lead to a somewhat reduced repertoire of possibilities.[2]

Situations of others are comprehended in presence: we put ourselves (with mobile consciousness) into the situation and co-experience the pressure on the person or character in this situation at this moment (as happens in co-experience, described below in the list of basic terms).

Predictive brain, prediction error. Neuroscientific measurements show that our brain is often slightly ahead of the external situation. When driving a car, for example, we react to the smallest stimuli that suggest that someone might overtake us from behind. We perceive a movement, not consciously, but we react by preparing the activation of muscles and possible actions. Our pulse increases. Emotions are activated. In most cases, this involves immediate expectations, and our brain produces specific expectations (predictions) about the most likely thing that will occur or has already begun. These predictions may or may not come true; errors may occur (prediction error), which are then corrected.[3]

Neuroscientists have summarized this idea in the concept of the predictive brain. They explain that such predictions are very useful, as they elevate us to a state of quick preparedness for an external situation. They also suggest that such predictions start a feedback loop that tests the prediction against incoming sensory information. While bicycling, we quickly predict that a bump could be in the road, and we start to compare the incoming information we are getting from our senses with this prediction: is this just a shadow or a bump that will require action? These predictions are ahead of our actual perceptions and to some degree we live and experience predictions, not perceptions, as Andy Clark suggests.[4]

It would certainly be wrong to call these minimal and barely conscious predictions narratives. Wrong predictions are usually adjusted without reaching awareness. However, we have to ask ourselves when the transition from immediate predictions to narrative-like expectations occurs. Consider the following case: you register a twitch in your partner's face, presumably without consciously noticing it. This twitch is in many cases associated with anger and rage, and somehow these emotions get registered. You adjust to this and suddenly you have the sentence in your head, "Maybe she will leave me." In response to this sentence, images and sequences appear in a flash, until you realize that it probably won't happen that way. Nevertheless, something sticks, a reality that is multiple in narrative terms now shapes our experience.

The point that I am hinting at here is twofold. First, our narrative-thinking predictions can become fully developed and fleshed-out forms of imagined futures. Second, some predictions stick and linger, even if they have not (yet) come to fruition.

Multiversionality. Stories are often exciting and can keep us spellbound. While we are in the world of a story and in a specific situation, we feel conflicting things simultaneously, such as suspense, disappointment, joy, fear, and hope. We have expectations and doubts and consider different possibilities. In the process, we end up mentally generating different, mutually exclusive versions of what might happen, or might have already happened. Maybe Harry Potter survives in the final book—but maybe not. When a friend tells us about his most recent date, many versions of what might have gone wrong pop into our heads. Talking about the model railroad in his basement was perhaps not the best topic for a successful first date. I call this consideration of different possibilities multiversionality.[5] Multiversionality is not an accidental or incidental property of narratives, as I will argue, but an essential feature of narrative thinking to make competing predictions and to mentally generate possible outcomes.

End, judgment, verdict. A central feature of narrative thought is that it occurs in episodes. The core feature of episodes is that they have an end. In this sense, Aristotle already articulated that stories have beginnings, middles, and ends. Although this sounds banal, it actually suggests that we actively seek an ending and, if necessary, invent one ourselves. The ending is important because it signals to our brain that we can now conclude the episode and devote our attention to something else. The ending thus accomplishes a number of things. First, it has the function of concluding a sequence of actions and experiences by fulfilling or disappointing the expectations we produced (. . . the boss started laughing).[6] Then it boxes up the entire episode, which can now be remembered as a whole in the form of a sequence with "first . . . then" (. . . he told his boss what he thought and stormed out. She then went to him and . . .). And finally, the ending also allows for an overall emotional appraisal, like a good or bad ending (. . . from then on, their relationship was much better). This emotional assessment is usually better remembered than any details or plot logic, as we will see. In this respect, the ending prepares the overall judgment or verdict.

Narrative, narrative thinking. This book will make a simple proposal of what marks the core basis of narrative thinking: narrative thinking happens in emotion episodes that entail the possibilities of different endings. Episodes have a beginning and an end. An emotion episode is an episode that ends in an emotion for the audience that typically has a rewarding function, such as a happy end. While the audience is engaged in narrative thinking, there are multiple possible developments, and without such imagined multiversionality, there would be no interest in the narrative.

Casting, casting of roles. When we dive into a new television series, we have to orient ourselves. We ask ourselves questions: Who are the main characters? What goals do these people have? What is the common mission connecting everything and everyone? As we orient ourselves, we place the characters into roles we know. We move to the roles complementing the leads, such as rivals, villains, antiheroes, and sidekicks. These roles guide us until we find our way in the story. When certain characters appear, we already know what to expect. One guy is the heroine's mentor, the other character is the sidekick, and so on. Particularly firmly anchored in our minds in this regard are dyadic patterns that relate two characters at a time: victim-perpetrator; victim-helper; hero-rival; mentor-student; lover-lover; friend-foe; giver-receiver; parasite and exploited, and so on.[7]

The repertoire we have for such roles and dyads is surprisingly small. When Vladimir Propp, the Russian folklorist, examined Russian magic fairy tales a hundred years ago, he found only seven different roles. However, it must be added that he also believed that all the fairy tales he collected were variants of one mega-story. The psychologist Daniel Wegner also attested that we get by with a small number of dyads to order our daily lives narratively. There are, of course, individual differences. Some people see the world sorted into perpetrator-victim roles; others prefer to recognize allies, admirers, or rivals in their environment.

In this book, the term *casting* functions as an alternative to the terms *schema, script, archetype,* or *stereotype.* Behind this is the assumption that we rarely grasp an entire narrative as a schema or script, because then the narrative would be too boring. Instead, we generate assessments of characters that indicate what we can expect from them in certain situations. But even then, everything can turn out differently.

Reception, communication. One of the wonderful properties of narratives is that we can pass them on to others. We can not only store our own memories as narrative episodes but also pass on highly complex contexts and events to others. This is not just about the facts of who did what to whom. Rather, it includes bringing about emotional reactions and making experiences. In respect to a range of features, the recipients of narratives seem to respond to narratives similarly to those who relay their own experiences, and they also remember stories in a similar way. The medium that determines what recipients of a story hear and what story producers want to convey is that of an experience. Stories convey an experience to the recipient.[8]

Co-experience, co-experiencing. The term *co-experience* is used in this book to describe the process of what a recipient undergoes when encountering the events and situations in a narrative. Recipients not only register the episodes, but mentally go through the action sequences at least to some extent as if they were themselves in the narrated situations. For this, different terms are used to emphasize certain aspects, such as *transport* (the recipient is transported into a narrative or media world, similar to mobile consciousness) or *immersion.* Co-experience is also a central aspect of narrative empathy: the recipient feels empathy for a character or can see herself in a situation as if she were a character.[9]

In particular, the notion of co-experiencing emphasizes the fact that we as recipients go through processes that can be reflected in our own experiences. That is, my friends' experiences can become, at least to some degree, my own. I don't have to make every experience myself. I can also communicate what happened to me to others. Apparently, we have learned to code our experiences in such a way that another person can match them to his own, thereby also expanding his experiences.

Groups of people that share stories are connected not only by a common vocabulary and shared frame of reference of myths and values but also by a reservoir of retrievable experiences and perspectives. Sharing experiences in the form of narratives creates connections with those around us that help us understand one another as we experience the world in similar patterns, roles, and sequences of actions. We live in narrative echo chambers.

Being trapped. We can be trapped in the wrong narrative. At the very least, we are often stuck in the patterns of narrative roles that we are most familiar with, such as perpetrator and victim, hero and rival, or mentor and student (part of the phenomenon of casting, described above). We build our mental narratives according to patterns that are familiar to us. This helps orientate us, but it also limits us and may reduce our perception of others whom we reduce to our expectations.

Psychologists like to use the term *schema* for such phenomena. *Schema* means something like a picture, outline, shadow, or basic idea. For temporal processes, psychologists use the term *script* to describe blueprints of expected actions and behaviors.

Narratives can create such scripts or expectations. However, narrative thinking also differs from such restricted schemas and scripts due to their open ends. When we process a narrative, we are in the midst of actions and sequences (the situation) before the overall sequence is known and resolved. The overall schema or script does not reveal itself until it is all over. Even if we guessed the correct script, we also likely entertained conflicting alternatives. In the middle of a narrative, be it the middle of a real-life episode or the middle of a movie, everything can still turn out differently. The prince can find me, Cinderella, and rescue me. Or I may run away in time before the fraudulent princeling appears. In short, every narrative always produces the possibility that everything can turn out differently (multiversionality). The trapping comes with a potential untrapping.

Cultural evolution. Charles Darwin described how living organisms could evolve over time and adjust to selection pressures. In a similar fashion, many cultural processes can also be described as forms of evolution. We can analyze, for example, how manufacturing knowledge about stone tools spread and improved during the Stone Age. In general, the incredible advances of our species involve a process where ideas, inventions, and ideologies are transmitted from generation to generation and evolve in the process.[10] It seems that our species has entered an age where cultural evolution determines our fate.

In this book, we are concerned mostly with a narrow focus of cultural evolution that shapes how narratives evolve over time. Narratives mostly offer non-technical knowledge, and their usefulness is less clear than that of a stone tool. Nevertheless, the spread and transmission of myths, stories, gossip, and all forms of narrative communication are likely a highly relevant aspect of culture. Myths, for example, represent a form of intellectual knowledge that can provide survival advantages to a group. They can entail warning information (don't trust strangers) and promote beneficial behavior (helping strangers can be good because they may help you). What would culture be without stories?

Richard Dawkins, in an apparent attempt to promote the idea of cultural evolution, emphasized its closeness to the Darwinian model by choosing the term *memes* for such cultural entities under evolutionary pressure. The term *memes* alludes to "genes" that nowadays are used to explain Darwinian evolution (Darwin, of course, did not know of modern genetics).[11] The suggestion is that certain memes are passed on more frequently than others in a struggle for survival of the fittest memes. In theory, this sounds good, but it should be emphasized that cultural units are much more formless than genes and, above all, can be passed on to anyone and everyone simply through communication, without presupposing complex reproductive processes and direct contact. Unlike genes, the presence of a particular meme in a culture may not be clear-cut, since memes are continuously modified as they are passed on. When novel words are used, when does a meme stop counting as the same meme? And what criteria are used to select the more short-lived internet memes in social media, for example, remains a mystery. Although some ideas survive and others disappear, it is unclear what the underlying laws are.

Instead of taking the notion of memes as a starting point, this book starts with the simpler hypothesis that narratives are adaptations. Narratives are passed on from one generation to the next and from one person to another, and each undergoes a process of adaptation when the recipients make the narrative their own. Put simply, people make them fit their world, their understanding, and their needs. By means of narratives, people try to explain their social world. People relay how others have acted, and this might lead them to consider how they themselves should behave and how that could be evaluated.

There is a second way in which evolutionary theory is important for narrative thinking. In contemporary evolutionary theory, the concept of *niche evolution* is somewhat controversial. The idea is that individuals modify their environment (their niche) and then pass on this ecological niche to their offspring who will have a better starting position since their environment or inheritance is already adapted to them.[12] No one doubts that there is evidence for this in the biological world, such as the beaver passing on the already modified lake to its descendants, but scientists disagree on how widespread this niche evolution is and what influence it has on evolution as a whole.

In cultural evolution, however, it is this niche evolution that is the very basis of development. One of the highly interesting features of the cultural evolution of narratives is that they shape the social environment and then this narrative-altered social environment is passed on. By sharing narratives, one mentally creates an environment for oneself and others and passes it on to the group.

A key concept for the niche we create is that of "culture": in our narratives, we create a culture and mental home with which we are familiar and, to some extent, comfortable. Culture is not a static entity we can simply point to; it exists by being evoked and by means of the narrative feedback loops that confirm, reinforce, and modify it. Such feedback loops also include the similar repertoire of roles and plot sequences that we noted above as casting, a feature shared by many cultural products, such as television series and movies.

As in all cultural forms that exhibit some stability over time, we can assume that they usually combine two aspects of adaptation. First, they seem to adapt to the cognitive mechanism of memory and communication. We pass on the information that better fits our brain. Second, they seem to offer people information that helps them in some way to deal with their environment. This could include providing stability simply because people are already familiar with these narratives and like to see them confirmed.

Brain. The approach of this book is to view the brain as a dynamic mediator among biological processes, brain activations, cultural contexts, and environmental pressures. This perspective, advocated by Thomas Fuchs, Rob Bobbice, and Adina Roskies, suggests that we can learn a lot about the brain by observing the dynamic interplay of the everyday tasks we face and the biological organ.[13] Put differently, our lived experiences are only partly shaped by elements such as brain architecture, biochemical processes, connectivity, and electrical impulses. They also integrate our situatedness in the world, our body, and its exposure to the environment, mediated by cultural dynamics. Put yet another way, our experiences cannot be translated into objective facts. They can, however, be communicated, and how we accomplish this amazing feat by means of narratives is the focus of this book.

My discussion of the brain will usually focus on the more narrow sense of currently measurable routines and processes of neuronal activations, as well as their temporal structures. Still, in my view, it would be misleading to draw conclusions about optimal narratives from, for example, brain architecture or the nature of synapses, or the way we convert sensory input into narratives.[14] This may be intriguing, but in most cases, this approach seems premature given our current state of limited knowledge. We simply know too little about the translation of processes in the brain to specific states of consciousness. We don't even know, and may never know, what exactly constitutes consciousness.

However, we can observe the extent to which specific brain regions are activated when people process narratives in movies, texts, or conversations. For example, in reading processes, we can observe when delays or more intense phases occur. Even here, of course, caution is needed because brain regions have individual plasticity in terms of their functions, and activation of a region need not be associated with the same function or even the same conscious content every time. There are many open questions in this rapidly developing field. When neuroscientists speak of a brain region, for instance, we should remind ourselves that as recently as 2016 researchers expanded the map of functionally differentiated regions from about eighty to one hundred eighty.[15] Some earlier studies are thus questionable since they simply equated regions that are close to each other but have since been differentiated.

To be sure, it would also be wrong to ignore how our biological brains adapt to storytelling and what we can learn from its operations as an organ. In this book, we will touch upon some of the ways our brains may help explain narrative thinking, such as the segmentation of experiences and how our prediction-making impacts narrative experience. For the related matter of our impressive reading abilities, see Maryanne Wolf, who suggests that reading is an act that does not come easy for our brains as they are not biologically well-prepared for this task.[16] I propose that this is different for narrative thinking, as our brains have evolved to encode experiences and to share them as stories.

Serial reproduction, telephone games, Chinese whisper. The term *serial reproduction* denotes a method from psychology in which participants are asked to repeat a story or a cultural artifact from memory many times—that is, to retell the same story repeatedly, or trace and retrace a picture over and over. In this process, either the same participant can reproduce the memory several times or chains of different participants each receive a story or artifact and pass it on to the next person, as in the telephone game (also called Chinese whisper). Serial reproduction experiments can be conducted in laboratories, but the process occurs naturally when artifacts of culture are passed on. In fact, it can be said that all culture is a large telephone game in which people pass on information from one generation to the next (cultural evolution). Myths and fairy tales are a product of serial reproduction, but so too are laws, religions, and many aspects of our daily life that have been passed on over centuries.

This method was popularized by the Cambridge psychologist Frederic Bartlett in 1932, and it has since been tested to study a wide range of phenomena, such as cultural evolution, memory, communication, social media, and art. Bartlett observed a tendency for memory to stabilize when narratives reached certain forms. He concluded from his experiments on the retelling of an obscure myth that this method was particularly well suited to distilling the basic forms of narratives and that this basic form consisted in their rational coherence, that is, in a narrative with causal links.[17] In this book, I will present a series of experiments to trace narrative thinking by means of serial reproduction. The startling result was surprisingly clear and quite different from what Bartlett thought.

Segmentation: Beginning and End

I'm not sure about you, but my life is pretty messy. When I wake up, I don't reliably know who I am or why I'm here. I may be holding on to some bits of my dreams, but even as they dissolve I'm not left with any certainty of what remains. Somehow, I find myself in the kitchen, magically drawn to my coffeemaker. By this time, I may wonder which other people should or could now be here with me, and I check some form of clock. I usually do this two or three times until I'm anchored in some sense of time. Does anyone need to be taken to school? Is someone waiting for me? Do I need to feed the cat? What emails did I not reply to last night? By this time, I've usually glanced at the index of several devices to check for incoming messages, but not opened them. They'll start nagging me soon enough. Instead, I've checked some news headlines. Some are about ongoing events displayed yesterday too, but they were somewhat different then. The wars in Ukraine and Israel, something has happened, and it may or may not be similar to previous events I read about. I know my hopes and I feel some sense of disappointment. Some news stories are fresh. I have never heard of this person but here is a picture of her in the major news outlets. While these images appear, images of other places come to my mind. Places where I do things. My lab. Also places I need to go today. Today? Go? Do what? Money or bills may come to mind, usually connected with an unpleasant sensation in my belly. And people. A person very dear to me is in big trouble. Suddenly this is fully clear to me. Her face pushes all other thoughts away. I recall recent moments of talking to her. Something bad has happened and now she's stuck. The situation calls for me to do something. Not sure what. But options may emerge. There is a sound; it's my little Italian coffeemaker on the stove. But that doesn't interrupt me now. I know I'm here to help my friend.

I certainly don't want to drag you into my world, but it helps to illustrate my point of being in a state of confusion for most of my day (which continues even after drinking my coffee). Focus and clarity are the exception. However, they are attainable. Retrospectively, I can talk about my morning in a clear fashion. But first, I need to develop a plan, decide what my target will be. In fact, the "me" seems to emerge along with a plan of what I will attend to. Planning and remembering what happened help me see connections, actions, and strings of events that survive fleeting moments of time. Once I can tell myself what I will do this day or how I will try to move my friend forward, I am somewhat together. I am someone, and I know what's part of my current "me" and my mission. What I am getting to is this: narrative thinking allows us to organize the messiness of our daily lives, beginning with the small, mundane tasks of every day and then on to the big challenges of life.

We don't need to consider the chaos of my head; we can go a little more highbrow. Let's consider this passage from the opening of Robert Musil's novel *The Man Without Qualities* (1930), one of the hallmarks of modernist writing:

> Automobiles shot out of deep, narrow streets into the shallows of bright squares. Dark clusters of pedestrians formed cloudlike strings. Where more powerful lines of speed cut across their casual haste they clotted up, then trickled on faster and, after a few oscillations, resumed their steady rhythm. Hundreds of noises wove themselves into a wiry texture of sound barbs protruding here and there, some smart edges running along it and subsiding again, with clear notes splintering off and dissipating.[1]

Musil captures the chaotic experience of traffic in Vienna in 1913 that defies clear beginnings and endings of specific events and instead creates a messy and convoluted totality. The perspective that Musil seems to take here is that of an outsider who comes from the countryside to this modern world. For this outsider, the challenge is tremendous.

The challenge that the modern city provided to Musil's outsider might resemble that of another newcomer to the world of senses: infants coming into the world. One of the great scientific puzzles is how infants manage to sort out the overwhelming variety of sensory impressions and orient themselves in their physical and social environment.

Luckily, we do not come into the world completely unprepared. Humans and presumably most vertebrates are born with a set of innate preferences. For example, we respond positively to the visual perception of faces and symmetry.[2] We respond immediately to some emotions expressed by others. Sound sequences we could already hear in the womb have high recognition value. But in many respects, we have to learn how to interact with the environment.

We must learn to understand our own bodily functions, including testing our sensory organs and refining our hearing to sort out the superimposition of different sounds and noises. Likewise, vision must be learned. Like the other great apes and many predators, humans use stereoscopic vision with a clear focus of attention on a narrow field of view. Thus, unlike many prey animals, which tend to have a panoramic view, we have a relatively small field of focused vision, but we have the advantage of strong depth perception and three-dimensionality. We are specialized in perceiving motion and quickly learn to follow a moving object with our eyes.

The next step is to understand the outside world. This world is full of objects, each with specific properties. They taste different. We can take some of these objects in our hands, but others are fixed and cannot be manipulated by us. Change and mobility also play a role here. A vase that falls from the table breaks. This is associated with a loud noise. Water also falls down, but it doesn't break; it drips and splashes. We get wet.

The world of an infant must be tremendously confusing, even leaving aside the difficulties of language acquisition and one's own physical coordination. Sounds overlap. Diverse sensory impressions attract attention. It is said that Dutch mothers, fearing sensory overload, placed their infants in cradles with closed curtains for this reason. We can see this in paintings from the so-called Dutch Golden Age of the seventeenth century. The idea still finds its echo in some practices today, such as infant education that emphasizes rest and sleep.

So how do infants learn to orient themselves in their environment?

Before different objects can be compared, an object must first be separated from the background and recognized as an independent object. This is a complex task. Two of my colleagues, Samantha Wood and Justin Wood, are conducting experiments with newborn chickens, which they are raising in a controlled world with virtual reality. They want to explore when an object that is perceived only digitally is recognized as an object. To do so, they are exploiting a characteristic of chickens: as animal researcher Konrad Lorenz has demonstrated many times, young chicks have the property of imprinting on one of the first things they see particularly clearly and assigning this object a positive or mother function. Lorenz showed that even his boots could serve such a function. In virtual rooms, these first impressions of the memorization phase can be controlled. It turns out that the chicks succeed in cognitively separating the virtual objects from the background when the objects appear to move.[3] However, not all motion is the same. There is consistent movement and there is seemingly erratic movement. A number of things belonging to the group of living creatures or robots have such seemingly unpredictable movement. They appear in the visual field, do something, and then disappear again. The momentum of these strange objects is also influenced by the behavior of the infant. When the infant cries, these things respond. Other things do not react.

What are the strategies available for infants to make sense of their environment? How can they make sense of their bodily and mental processes? On the most basic level, sense making involves drawing lines and making distinctions.[4] This is where one entity ends and a new one begins. We classify entities and segment temporal processes.

As a first start of orientation, infants and children receive many important impulses from their mothers.[5] There are welcome, neutral, and unwelcome things. Objects can also be divided into groups based on similarities, which can be distinguished from one another. The group of liquid things behaves differently from the group of hard things. Things from the group of living things behave differently than things from the group of robots. Soft things make different sounds than hard things when they fall. Such classifications, once acquired, allow orientation.

For this book, temporal processes are of special importance. Temporal processes can be segmented. They have a beginning and an end. Segmentation is how we make distinctions in the temporal world. Once a segment is cut off from other processes it starts to represent a small unit or episode. A thing that is thrown first flies up and then down until it hits the ground. Then it lies still and stays in place. This sequence is a minimal unit. Wood is piled up; someone kneels in front of it until it starts to become bright and warm, and flames can be seen. A noise outside the door announces that the door is about to open and someone will appear. The segmentation of temporal processes allows us to see what belongs together. The sound at the door signals a person at the door; that in turn signals an arrival. Piles of wood belong with fire and warmth. We cut out small episodes from the many overlapping actions and events, observing that they usually occur together. The temporal proximity allows for predictions of what will happen when one of the elements of a sequence is perceived. The sound of the microwave suggests food. The accumulation of wood in the cave shows that there will soon be heat. Someone is coming. The world becomes more predictable. Temporal processes also attract attention. As the studies with chicks show, chicks pay special attention to objects that move. One may perhaps assume that objects only become recognizable as objects for humans, too, because they appear in temporal chains in which they can be detached from the background.

The formation of temporal units represents a fundamental principle with which humans and many other animals understand and order their environment. The segmentation and the bundling into small temporal units make experiences rememberable, manageable, and applicable in many ways. At some later stage, segmentation also allows sharing, co-experience. The temporal units or segments fall within a spectrum from the general to the particular. Some insights have the status of general knowledge that is characterized by a when-then. When it gets cold, then water freezes. These linkages of related sequences operate largely independently of one's own experience. Those who have learned these linkages can rely on them and expect them automatically. Only when something else happens do we react with surprise and become aware of our expectation. Babies laugh, for example, when a balloon does not fall to the ground according to expectation but rises up.

Other temporal units or segments are one-time events. Yesterday, the fir tree in the garden fell down. My friend's news put me in a state of great excitement. These unique events are experiences in the true sense of the word, since they are based on a specific observation at a specific point in time or, in the case of social events, on a co-experience. To distinguish general when-then events from one-time events, biologists often distinguish between semantic and episodic memory. Semantic memories collect general knowledge about the world, while episodic memories denote one's own, one-time experiences.[6]

It's a matter of debate whether non-human animals also have episodic memory or only quasi-episodic knowledge. Endel Tulving, who coined the term *episodic memory*, reserved it exclusively for humans, who have free and conscious access to their memory and can recall it at will.[7] Episodic remembering allows "time travel" or what I call mobility of consciousness. Of course, non-human animals can also store individual events and learn from them. Rats, for example, have a very accurate memory with respect to single experiences, like remembering a maze. They can also recognize various conditions that must be met for one single experience to resemble another. For example, rats can learn and remember that a door in a maze leads to chocolate only if they have not been there for a long time.[8] In this process, the (quasi-episodic) single experience is transformed into general (semantic) knowledge. With Tulving one could formulate that the single experience in this case possibly exists only with respect to its possible semantic generalization.

These differences between individual personal experiences and general sequences are important for us for several reasons. First of all, they raise questions about the agents or subjects of an experience. Do narrative episodes require an agent with a mind who can have episodic memory like "At that time and place, I decided to jump on the ice . . . "? Or would an inanimate process satisfy our expectations: "Snow melts, some rock on the snow starts sliding down the mountain, and then splashes into a newly formed lake"? While we can certainly agree that inanimate processes such as the sliding rock can be segmented into closed units, there is also something unique about events that involve a being with a mind. Beings that have an inner life and perform self-directed actions, such as humans, animals, robots, and mythical creatures, act according to principles that are usually not fully apparent to the observer and thus attract special attention.[9]

More generally, Tulving's difference between episodic memory and more general, semantic memory allows us to raise the question of how the unity of an episode comes about, that is, its *episodicity*. It turns out that the clear distinction between semantic and episodic memory does not capture human experience well. We seem to be able to shift between them. What about the events that one has empathetically experienced? Do these memories now denote one's own (episodic) experiences or are they (semantic) knowledge about the (social) world of how others in general experience events corresponding to some folk psychology? When I hear that a colleague's old dog has died, I might vicariously experience this news or I may focus on the general pattern of sadness that tends to emerge from such events.

The same holds also in the reverse direction: we can recognize the general (semantic) structure in our personal and unique episodic memories and extend unique experiences into generalizable semantic knowledge. For example, a student asks me to put in a good word for her to receive a fellowship. I state that I am happy to comply and write a letter of recommendation. I start to wonder how I can make a case for this individual. During that process, I notice that my feelings have become quite positive as I care for the student's future and perhaps also feel honored by her trust in me. . . . I could now easily get carried away in writing an overly positive letter. However, at that point I realize that I am experiencing something close to the Ben Franklin effect, where people are doing good things for people for whom they have done good things before (in this case, I first agree to help and then, as a second step, write overly warmly). Now, once I realize my reaction, I start to observe a pattern. My feelings and thoughts shift from the individual student toward some general awareness of people and my task to write a fair letter.

Maybe that's a bad example, too peculiar to the odd world of professors. However, the idea here is that we can become aware that specific episodes and specific understandings of behavior always also show some general pattern or bias of behavior that we come to expect. Thus, our personal episodic memory is porous for semantic understandings; that means it involves both unique features of the one-time event and general patterns of understanding that can start to overlap the unique memory. Perhaps a distinctive feature of human memory is that humans flexibly manage the boundary between the two. We can extend our "own" unique experiences to those of other beings, including fictional ones, and seek the general semantic pattern involved.

We are getting ahead of ourselves, but narratives might let us accomplish this task. Individual experiences can be shared with others. The unique episodic experience becomes a little bit more "semantic" in the process. Narratives operate as a medium between people and allow co-experience. In the process of creating a narrative, the radically unique experience and memory become a little bit more generalizable.

For now, however, we will address the question of how we form these unique temporal units, or episodes, and how they structure our experience. Neuroscientific studies show that people are particularly attentive and invest mental "energy" at the beginning and end of observed short episodes. These studies on event segmentation give us a number of interesting insights and are one of the most important contributions of neuroscience to the mystery of narrative thinking so far.[10]

Segmentation structures the constant flow of life. The clearest boundaries are always drawn when human actions change. What could possibly seem chaotic is transferred into a minimal order. Beginning, then comes an end. Whoever segments more can remember better.[11] The implication is this: whoever segments more is thereby "smarter." Whether this is true for TikTok users who are provided with segmentation of short episodes remains to be seen. But it is obvious that the short clips in TikTok take advantage of our preference for fast segmentation.

People can recognize and isolate everyday action sequences, such as washing one's hands or making coffee, thereby turning them into an episode composed of numerous small individual actions. We can also create more complex sequences. When people watch a movie with complex social interactions, most agree on where they place the boundaries between episodes. They also show higher degrees of concentration at the boundaries between different episodes. People also read more slowly at the transition points from one episode to the next.[12]

It turns out that good segmentation allows better memory. When past episodes are recalled, the entire event space becomes accessible again. In the process, errors and false memories may occur, as only what seems to fit is added. Perhaps consciousness is composed of such episodes.[13]

Before drawing conclusions about the possible consequences of segmentation for narratives and narrative thinking, we should note that segmenting episodes is, cognitively speaking, particularly costly for our energy budget. If this were not the case, the measurements of brain activity would not be able to detect it so clearly.[14] Thus, it can be assumed that the segmentation must be advantageous, but what exactly is this advantage?

To answer this question, we need to gain clarity about what is segmented. How is separation of continuous time achieved? One clue lies in the spatial metaphor of event "borders." Events obviously do not have actual borders, and we borrow this metaphor from spatial concepts. Now, actual spatial boundaries do play a role in segmentations. In memory test experiments participants were asked to move from one space to another. For example, some walked through a door, others did not. As a result, the two groups recalled the items on the test differently. Those who passed through a boundary generally had a less accurate recollection of the items than those who remained in front of the boundary. This effect was evident in both physical spaces and virtual spaces in computer games, as well as in spaces that were narratively created and communicated. Spaces produce the effect of proximity. What is in a room is classified as belonging and is remembered as such. Three things in one room are remembered better than three things in three different rooms. Neuroscientists assume that the mental model of a situation is "updated" at each of these boundaries—that is, not only refreshed, but also reloaded.[15] As humans, we are mentally only in one situation at a time, and it is correspondingly important to draw the boundaries between situations or event spaces. Out of sight, out of mind.

People make better predictions within an episode or within a situation compared to predictions across situations.[16] One option to explain this tendency could be to suggest that the possible events within an episode are understood as more coherent or appropriate for the situation. Another explanation would be that we pay more attention to the objects and options within one situation or event space, and thus predictability is not an effect of increased coherence but just attention. We may also explain the better predictions within one episode in reverse: the clearly predictable events belong in one event space, while the less predictable events are shunted off to another event space and thus marked by a boundary.[17] Event boundaries mark the event spaces about which one currently has knowledge, while the uncertain is exported to the outside.

We are now arriving at the insight that event segmentation is a significant cognitive accomplishment. Segmentation draws our attention to the immediate concerns of a situation. It can thereby raise our awareness that some actions are not yet completed and keep us prepared for their imminency. Once they are completed, we can understand the episode as closed.

In many cases, the lines of what functions as the beginning or end of an episode are not clear or given, and the drawing of the line is itself a cognitive act and decision. There are short stories or episodes that people view in different ways, where we can draw the lines of beginning and end differently. Take, for example, the famous controversy or divisiveness of Shel Silverstein's children's book *The Giving Tree* (1964).[18] In this story, a tree keeps giving apples, shade, or wood to a growing boy until he turns into an old man. Some read the story as a beautiful example of a happy parental giving situation, while others condemn the eternal boy as a selfish brat without empathy. The story operates a bit like the famous rabbit-duck illusion that Ludwig Wittgenstein reported on, an image where people see either a duck or a rabbit. In the case of *The Giving Tree*, people may see the ending frame as parental fulfillment of having given everything to the child, or they may see the story as unfinished with a boy who never grew up and never learned to appreciate the tree. The end thus would occur only beyond the story.

It turns out that many episodes in our lives have more features of *The Giving Tree* and the rabbit-duck illusion than of the simple closed segments like washing one's hands. When I interact with people, I often do not know whether some exchange has come to an end or is ongoing, or whether I feel like this exchange ended happily or requires some further negotiations. A friend tells me at a coffee break that life sucks, and I try to cheer her up, but we both need to move on. This is not a closed episode, but it may rest nevertheless. We watch as a student sneaks into a professor's unattended office and, after a short time, walks out the door with a laptop. The episode is finished, but of course nothing is finished in our minds. We do not know whether the student acted on behalf of the professor to fetch the laptop or whether we just witnessed an act of theft. Should we intervene? If so, how, without making the student—likely the professor's assistant—feel awkward?

In general, what constitutes the closedness of an event space? In many everyday cases, we are dealing with spatial units of events at a place and time. In these instances, the segmentation here is really at the boundary or passage from one place to the next, usually on the basis of very small periods of time. Some other units are certainly drawn by strong associations of belonging, such as the setting of a table, which, although often done between two rooms such as the kitchen and the dining room, is held together by the dishes and the functional actions. So far so good.

Beyond the basic everyday actions, things become messy. Think again about my unstructured mornings or Musil's cityscape. Many events do not lead to a clear end. There are actions started that lead to several consequences, but none of them close the starting action. I issue a complaint, it triggers reactions, but when is the case closed? Several people meet to brainstorm what could be done to reduce homelessness, but is this brainstorming event over when people get up and leave the cozy room?

In many such cases, it becomes apparent that the end is not immediately part of event spaces or segmented short episodes. A person pushes forward to make a career by elbowing our friend out of the way; the act is complete, the promotion is accomplished at the cost of our friend, yet we wait to see if there is a later punishment for it or not. Someone is in love, glances are exchanged, but no one knows what will happen in the coming days. Readers of fiction may recall *The Count of Monte Cristo* (1844), by Alexandre Dumas, in which the protagonist finds himself on a prison island that puts Alcatraz to shame. He is innocent and was sent here by his rivals. Revenge drives him and his readers along in hope of the impossible escape. As readers, we are guided by hoping for his big comeback and wait for the comeuppance, and we do not consider the novel complete until that point.

In such cases there are expectations resulting from the actions and our hopes, but they are not fulfilled in a single episode. Obviously, our interpretations of what happens in an episode play a big role in driving the expectation. We do intellectual work to judge an episode as completed or unfinished. That is, to some degree, we *decide* to count an episode as finished.

We can learn to see and declare an episode as closed. When participants in an experiment are presented with random combinations of elements, they can, with some training, learn to recognize each as an event space with boundaries and to wait accordingly for repetitions of similar patterns.[19] Memory artists often report that they arrange random patterns into strings of primitive stories with beginnings and ends that they can store and recall better. We can code arbitrary combinations of elements in such a way that we see them as the proper ending of a string. Closedness and openness can apparently be learned and then play an important role in our mental orientation.

Learning to close segments has its advantages. However, we typically don't designate ending points arbitrarily. We don't just compartmentalize our lives without thoughtful consideration of what transpired. In the realm of narratives, multiple layers of hierarchical organization likely contribute to organizing segmentation. Apparently, people recognize similar event boundaries in narrative sequences when they reread them, and they skillfully orient themselves to them.[20]

Different people encode similar stories in a progressively similar manner, as revealed by brain imaging studies. This similarity extends to patterns of reinstating past information and integrating it into complex, ongoing narratives. The segmentation of episodes appears to facilitate the hierarchical organization of key elements in ongoing stories, such as protagonists and key motifs, into an overarching structure. Perhaps this organization could be described less as a hierarchical processing but more like a "stitching" together of episodes.[21]

The lingering question revolves around how these hierarchical event boundaries are established in the context of narratives. When does an episode appear to us as "resolved" and when is it "unresolved" and not just factually "finished"? This question goes beyond the question of physical spaces and event spaces. One of the answers that neuroscientists and psychologists have given to this question consists in immediate causality.[22] That means an episode is closed when we understand what caused it. The cookie disappeared because Cookie Monster got hold of it. Still, if we were to generalize this idea of causality, it becomes somewhat unsatisfactory. It is true that an episode proves to be somehow "finished" when a causal sequence comes to a conclusion, but that does not mean that everything is now over. Snow White is abandoned in the forest because her stepmother wants her killed, but she is then rescued *because* the huntsman takes pity on her and lets her go. There is causality here, but it is not an ending. The roommate in the shared apartment gets drunk and smashes up the kitchen. The plot is finished and has causality, but it is not finished, either in terms of consequences or in terms of how it came to be. So how are our expectations of narratives controlled, and when do we consider a narrative resolved and concluded?

We can draw two simple and preliminary conclusions so far. The first is that segmentation is decisive for how we think in terms of narratives, but the second is that the borders alone do not explain what makes an episode. So far, we have focused on the borders of episodes, the beginning and end, that interrupt the continuous flow of time. Segmenting the temporal flow allows us to process temporal units we can handle and remember well. Order and orientation are possible. However, it seems we are missing the bigger picture of what makes an episode complete. Not every end of an episode is satisfying and operates as a clear end to us. Something can be over without having ended. The point of these considerations is thus to tell us that we do not know enough about narrative events when we merely understand the event borders. To get a more complete understanding of episodicity we need to consider what happens in the middle.

The Importance of the Middle

Is this it? Do we sort the world in stories simply because they can be chunked with a beginning and end? Certainly, as we have seen, the encoding of beginnings and endings is backed up by evidence from brain studies. However, it would also be rather disappointing if this "chunkability" is all we need to know about narrative thinking. Consider these two mini-stories:

> Hansel and Gretel were two children who lived in poverty with their father, who loved them very much, and a stepmother who did not. Lots of things happened, and at the end the kids and their father lived happily without the stepmother and were rich.

> When I was ten years old we were on a vacation in Sweden in a remote area. It was a long time ago, before we had cell phones and email. My father got a telegram that he had to pick up a document (he worked for the ministry of defense, this kind of thing happened). On the way back, he stopped at an isolated mountainous area, but he never came back. A large search was started. His body was found several days later. We were told that it was a mountaineering accident. I only learned recently that this was not true.

These mini-stories have a beginning and an end—but they are also utterly frustrating and disappointing. It is not just information we are lacking. Sure, we're left wondering what happened and how the beginning and ending fit together. But more important, we're robbed of the very point of the story that makes it worth telling in the first place. The stories withhold from us not only what happened but the reason why we should care about what happened. Without the middle, they are almost pointless. (Of course, we may get inspired to fill the missing middle ourselves, but that would only prove the point that something is missing here.) We will return to both of these stories later. The second story, the story of my life, in particular is interwoven with this book as I learned the truth about my childhood only recently while I was writing this book.

So, what about the middle part? Let's recap. The theories and studies of minimal episodes discussed above emphasize the borders—that is, the beginning and the ending of the stories. By means of these borders, episodes can be detached from their context and isolated. Narratives proceed in very different ways, but they all begin and end somewhere. The ending, in particular, serves as an anchor. A narrative ending is usually very clear. Something is finished—or obviously unfinished, as in a cliffhanger. The ending can concern simple actions, but also complex relationships. In tragedies, the ending is often associated with the death or closure of the scope of action of central characters. The happy ending often relieves some tensions or comes after some emotional flow of prior negative emotions.[23] There can also be the satisfaction of punishing bad guys. What these endings have in common is that they are usually preceded by a period of that very ending being delayed, reinforced perhaps by anticipation and wishful thinking. Perhaps there is no happy ending without such a delay.[24] But as the two examples of stories that lack the middle part show, we do not yet know what constitutes an episode, except that it starts somewhere and ends somewhere. We do not actually know why the borders are where we draw them.

In this section, the accent will now be directed to the middle of the narrative arc. If we were to base this investigation on our observations of event segmentation, the middle would be determined merely as what happens between the beginning and the end. Based on the minimal episodes and the neuroscience studies discussed in the preceding section, one might be tempted to explain this middle not as a moment in its own right but, for instance, simply as the delay and expectation of an ending. After all, the neuroscience studies found no clear pattern of arousal in the middle. So the question arises: how and with what vocabulary can one describe the development between the beginning and the end of a story? Does something decisive actually take place in the middle?

When we ask about what happens in the middle of a narrative episode, we need to start considering the big challenge of defining the human experience. However, we do not have to jump there in a single bound. It helps to keep some careful distance and to follow methods and established paths of reasoning. For that reason, it never hurts to go to the classics for inspiration. In the following, we will first engage in a discussion of two classic understandings of narrative design. This will then lead us to a first potential explanation of how narrative episodes make other people readable to us and available for co-experience.

Aristotle set forth the notion of *peripeteia* for what takes place in the middle. This is the turn or reversal of a path from which the narrative swings from one course to another before it can come to an end. He states in his *Poetics*, "Peripeteia is . . . the transformation of the plot into its opposite, and that, as we say, according to probability or necessity."[25] For what such an "opposite" can be, Aristotle gives some examples, such as the sudden change from happiness to unhappiness and the change from not knowing to knowing that occurs in recognition scenes. What is crucial for Aristotle is that this turnaround involves sudden changes in mental states, like the sudden realization of knowledge or the quick emergence of happiness. In the case of a deus ex machina, the emphasis is less on the actions and events accompanying the arrival of the rescuer than on the hopes associated with the rescue and thus on the mental states of protagonists or spectators.

We are thus getting two clues from Aristotle. The middle is about change, and the middle concerns our own affects as audience members of the drama. For him, this change is more of a mental activity by the audience than a plot twist. In scenes of recognition, when one character suddenly recognizes a long-lost family member, the facts do not change but rather their mental representation and corresponding affective evaluation do.

In the nineteenth century, the German writer Gustav Freytag, on whom we will draw in the following, offered an interpretation of Aristotle's ideas that has influenced many generations of writers and scholars up to today.[26] Like Aristotle, Freytag emphasizes the middle with the change and reversal. Literary scholars and creative writers in

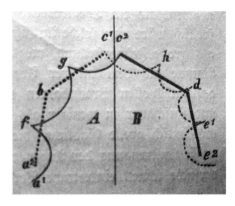

Fig. 1. Freytag's depiction of Schiller's *Wallenstein* drama with rising and falling developments (Gustav Freytag, *Die Technik des Dramas*, note 19)

writing workshops nowadays remember Freytag's theory as "Freytag's pyramid." It is worth taking a look at. Of course, as with Aristotle, we are initially dealing with works of art and fiction. But we can hope that here, too, we might find insights into the less exceptional everyday narratives.

Freytag's pyramid is famous above all for its clear schema of rise and fall. Interestingly, Freytag contrasts two models as equivalent: on one hand a pyramid or reversed U-shape with rise and fall, and on the other hand an upside-down pyramid or V-shaped figure with fall and rise. For both of these shapes, Freytag distinguishes eight phases of this development in five parts and three places, which can sometimes be further subdivided (fig. 1).[27]

What is decisive about Freytag's figure, however, is not so much the sorting of the phases but the rarely discussed preliminary remark that puts the psychological moment of this figure in the center. Freytag emphasizes what actually distinguishes the ascent and descent: in one case, the protagonist is the active part of the movement, and in the other he is the passive or suffering agent. The climax and turning point thus mark the point at which the protagonist's participation changes from passive to active or active to passive.[28] It is precisely this change that marks the middle of the narrative arc.

In Freytag's words, the "contrast" between active and passive consists in "one's own deed and the deed of others."[29] He also speaks of a contrast of inside and outside. To describe the active dimension, Freytag speaks of a will that acts "from the inside out." To describe the passive dimension, he speaks of the reverse movement that emanates from the "outside world" that "rises above" the protagonist and impacts him (from the outside to the inside). In the latter case, the arc begins with the hero as a "recipient, sufferer, who is overwhelmingly determined by the counterplay that arises from the outside and strikes him." Freytag emphasizes how the active and passive dimensions are directly connected and show a reversal of the balance of power. When the movement leads from active to passive it means that "what he has done comes back to affect him." In the case of the reverse movement from passive to active, the narrative arc begins with the "alien forces" gaining an "influence" on the protagonist's "interior" until the protagonist resists and reacts. The crucial thing about Freytag's hypothesis here is that the intensity of one turns into the strength of the other. The stronger either the active or passive forces are, the stronger the flip side of passive suffering or active struggle will be.[30]

According to Freytag, a narration is successful only when both dimensions of active action and passive experience are in a state of balance and occur one after the other. It is not enough to expose a protagonist to a blow of fate or to stage her as a victim. She must also be shown as having acted. Even the failure of an action alone, according to Freytag, does not constitute a closed drama and tragedy, because the inner emotional reaction of the previously acting person—that is, the passive suffering—is still missing.

Although this standard was developed for high literary fiction, many everyday stories may actually live up to this claim and Freytag's pyramid as well. A friend has been suffering from his working conditions for years. He has reached a dead end in his company with no opportunities for advancement. Although he could have a quiet life there, it was not what he expected from life. We observe this, but, having become accustomed to it, we hardly notice it anymore. Suddenly, he gets excited and wants more from life and from his job. Now we are fully involved, want to know what he can do, and experience his adventures in the company and at job interviews. The passive friend has become an actor who rebels. And we as friends experience it, he has our full attention, and we are committed to provide support. He has a story.

So, what is the reason for this switch between active and passive experience of the protagonist? Why does this switch lead to an intensified perception of an episode with a beginning and an end? To understand this, we need to emphasize an aspect that so far has been only implicitly in play. What has been unstated, but central, is that this sequence of active action and passive experience takes place in *the eyes of the observer*. (Beyond the metaphor of the eye, we should emphasize that it takes place in the entire body of the observer, because it is in the body of the co-experiencer that tensions are generated, feelings experienced, and expectations worked through.) It is the spectator in the theater, the recipient of the latest gossip, or the observer on the street who witnesses the character's turn from active to passive and who connects the behavior and experience.

We constantly observe others. We do this in daily life when we track people we know but also when we take note of others we do not know well or at all. We also track characters in television shows, the theater, and fiction. The goal of such scrutiny is to catch that which cannot be observed—namely, the inner being, the mind, and motivation of the other. Observers consider the actions of others and thereby hope to deduce their inner states—their thoughts, intentions, and feelings. To emphasize again, this assessment is quite complex, as the inner states of others are not directly accessible to observation. However, we have one vehicle that guides us: emotions. Emotions give us access to others; they transcend the boundaries from the inside to the outside and from one person to another.

This challenge of observation is at stake when Freytag describes the passive experience of the protagonist as the emotions of the protagonist. As long as the protagonist engages in action, we, the observers, might watch the protagonist merely from the outside. This changes in the case of the protagonist's inner and passive experience when the struggles become manifest. Here, the protagonist feels and shows inner states and emotions. These emotions put the protagonist's inner life in front of us, make it a fact, which in many cases will lead the audience into taking an insider or first-person perspective. That is, the transition from active to passive or passive to active often corresponds to a switch of perspective by the audience. Put differently, the move from one perspective to another opens up a space for navigating how we can relate to the other person or character.

In this sense, the eminent researcher Nico Frijda describes what the concept of emotion is supposed to accomplish—that is, a probing of the discrepancies between the external and internal worlds, such as between "what people do or feel and the events around them."[31] According to Frijda, emotions mark the distance between the external world and inner states. If we are unhappy, for example, this would mean that we find ourselves in a situation that is not as we would like it to be. When we perceive this tension of inner state and situation, we experience it, at least to some extent, from the perspective and the body of the protagonist.

We should note here that the standard theory of emotion in psychology, the appraisal theory of emotion, takes a similar starting point. It suggests that we evaluate or appraise a situation and thereby create an emotional or affective response to the situation. When we experience a situation, we evaluate or "appraise" it, and this appraisal manifests itself in an emotion. Interestingly, this appraisal also offers a bridge to other people who are not in the situation, as they may ask themselves, How would I feel in this situation? To be sure, such appraisals are not the same as empathy. We are not feeling what a distinct other being is feeling. Still, this understanding of emotions as appraisals of situations allows some transfer. The other person in a situation can become an avatar of our self since we might both feel the same in those circumstances. The discrepancy of inside and outside manifests itself as a kind of loop for observers as they explore the difference between the internal being and the situation.

The key move in Freytag's theory of the pyramid is thus not simply that the energy of active actions can be transformed in passive suffering or vice versa, but rather that the external situation and the inner feeling can be mapped onto each other. The situation can make the mysterious inside of a character readable—and this does not just mean understandable but also available for co-experience. In principle, we as humans can read and co-experience other people when we only have access to one or the other state—that is, when we observe or know of their inner experience in a situation or we witness their agency. However, the narrative design of the pyramid becomes a powerful tool to focus the observer or audience on the experience of the other. The distance between the active behavior and the passive indulging of the external situation becomes the bandwidth within which the audience can estimate the state of the other person or a protagonist in a drama. The flip from active to passive or passive to active triggers the audience to estimate how far the actions that follow the passive state will go or how deep the suffering following the phase of agency can be. Thereby, the audience takes a position or perspective at least partly from within the other person to contemplate that person's feelings and actions.

This switch to a perspective from within the other person might be the key to understanding the middle of a narrative episode. People possess a "mobile consciousness" with which they can put themselves in the most diverse situations, and this mobile consciousness reacts to triggers and attractors. Narratives often operate as these attractors or triggers. Co-experiencing the situation of others is a particularly clear case of such mobility of consciousness. It therefore seems to be a reasonable guess that the events in the middle of an episode play a role in directing this mobility of consciousness.

Both the training of mobile consciousness and the withdrawal from co-experiencing will occupy us later in this book. Crucially, the observer's attention demands not only a beginning and an end to an episode, but also something that draws him into the episode and allows him to co-experience it. This could be the shift from active to passive or passive to active that Freytag describes. One can also put it this way: in a narrative episode, the recipients of the narration are spellbound and captivated because they want to know what the narration is leading up to. Only at the end of the episode are they released. Only there do the expectations come to an end.

Accordingly, a minimal narrative or episode does not only have the boundaries of beginning and end but also this turning of the protagonist's experience, which can be witnessed from an observer's perspective. This turning point connects the beginning and the end, but it also causes a *change* from the beginning to the end.[32] According to Freytag, this turn represents the climax and turning point of a narration and marks the moment when the protagonist—and with him the observer—is drawn into the narrative.

In a similar sense, Aristotle spoke of catharsis as the climax of tragedy. Catharsis is the moment of supreme witnessing, but it is also, according to the Greek meaning of the word, a purification, a detachment of the spectator from identification with the character. It is precisely this return to oneself that is promised with every narration, and because we can expect narratives to come to an end, we engage with them more willingly. Narratives abduct and hold us, but we only allow this because we know that the hostage-taking will end. This is the contract that every work of fiction makes with us. (Some scholars question whether the modernist literary works that attempt to deny such an ending break this contract, but that issue need not be decided here.)

At this early stage of our investigation, we have used ideas from Aristotle and Freytag to shine some light on the middle of episodes. This has led us to the idea of the turning point, or peripeteia. We considered the turning point as the place where the emotions and inner states of the character of the story appear more clearly to observers and can be registered by them. We theorized that the turning between active and passive modes of the character might be especially well suited to attract attention by observers who are then drawn to take the perspective of the character.

Let me add one final note that will turn out to be more important later. The co-experiencing of a character's inner states typically conveys a positive feeling of presence even with negative emotions—*for the observer*. The person who feels an emotion will experience it as either positive or negative according to what the emotion is, but observers may bend it toward the positive, since it evokes this sense of presence.[33] That is, observers in their co-experience can also gain something from the negative emotions, or at least this is what Freytag implies.

45

What Have We Learned?

Narrative thinking is oriented around narrative episodes. Episodes interrupt or gather the flow of perception into segments. These segments allow us to process temporal units and event spaces. Segmentation enables better orientation and memory. Neuroscientific studies show that the delimitation of episodes takes a lot of effort in the brain. Apparently, the cognitive gain of orientation is worth the investment.

An episode has a beginning and an end. Episodes also have a middle. In an episode, protagonists act with an internal motivation to achieve a goal, or they react to external circumstances. In the course of the episode, the protagonists can appear both as active agents and as passive reactors. According to some aesthetic theories, the middle of the episode represents the transition from one to the other. This transition emphasizes the emotions and inner states of the protagonists, thereby inviting observers to co-experience them. The episode is over when this transition from one to the other has taken place, because then the co-experiencing observer or recipient can withdraw.

When I was about to take the oral part of the official Graecum exam at university, I breathed a sigh of relief when I read that the topic was to be Aristotle's *Poetics*. I was prepared for this. When asked what Aristotle said a tragedy was and how to recognize it, I was able to explicate immediately the common interpretations of the *Poetics* current at the time. But I soon realized that I had not convinced some of my examiners. On the contrary, they indignantly lectured me that, according to Aristotle, a tragedy is a work with episodes that has a beginning, middle, and end. Period. At the time, I felt that the lecture I had received was pedantry. Now I recognize the wisdom in it.

two

TELEPHONE GAMES

In July 1870, the French ambassador, Vincent Benedetti, was instructed to meet the Prussian king. He met the king at his vacation spot in Bad Ems during a walk on July 13th. There he made demands on behalf of Napoleon III that Prussia should state it would abstain from interfering with the selection of the successor to the Spanish throne for the indefinite future. By the accounts we have, the Prussian king firmly, but not in an overly confrontational way, refused to give in. That encounter led to a chain of communications. First, the king wrote to an administrator. The administrator then summarized the events and wrote to the Prussian chancellor, Bismarck. Still on July 13, Bismarck edited and rewrote the letter he received and released it to the press and other diplomats. The German news text was translated into French for the French press on Bastille Day, July 14. At each stage of the transmission and translation of the so-called Ems Dispatch, changes occurred that accentuated various aspects of the confrontation. The communicators were all highly influenced by their nationalist and patriotic emotions. Their perspective shaped how they saw the message, and they used the communication to evoke and confirm the very emotions that would fit their understanding of the situation, even though they each might have claimed and thought that they passed on a neutral description of what happened in the encounter. Apparently, both sides found plenty of reasons to feel anger and rage about the other side. The rest is history. On July 19, 1870, France declared war. The war resulted in a humiliating defeat for the French, paving the way for German unification in 1871 and further German-French animosities that shaped the twentieth century with its world wars.

The process that resulted from the famous Ems Telegram (or Ems Dispatch) imitates the proverbial butterfly. In the case of the butterfly, the original movement of air turns into a storm on the other side of the world. In the case of the Ems Telegram, the communicators turned the disagreement of the original message into full hostility. In both cases, the process of transmission brings out something that was intrinsic to the original but presents it in a much clearer and amplified form.

In this chapter, we turn to a specific method for observing narrative thinking in action—namely, chains of communication and, more specifically, the retelling of stories in the form of the telephone game. Why retelling? Retelling reveals what has taken root in a narrator's mind and how she remembers, summarizes, and makes sense of events. Retelling allows us to compare what is dropped and what is accentuated between the original and the retelling. Academics may feel tempted to speak of the retelling as an "interpretation" of the original story, but this academic jargon seems to imply some intellectual process of hard thought. Instead, retelling happens easily and naturally. When we retell a story, we may "see" the original events in our inner eye like a movie or go through the plotline as if we were experiencing it again. When we retell an event that we witnessed or heard about, we create it in our minds without fully noticing we are creating it. The parties who reiterated the episode of the Ems Dispatch probably all imagined the scene and then each told it in their own words.

This activity provides fabulous opportunities for research. First, we catch the act of creating a story in the act. Second, we can compare the original and the retelling and consider all changes. And third, we can manipulate the original stories experimentally to see what changes each change triggers. Yoshihisa Kashima and his collaborators have shown, for example, that information that is consistent with stereotypes is maintained better and exaggerated compared with other information.[1]

From the standpoint of the reteller, there are combinations of processes that connect reproduction with creation and adjustment. First, the retelling shows how the recipients have received and understood the story. Second, there are memory processes, which lead partly to forgetting but also to embellishing and adjusting to the memory process. Third, the ideas that the reteller has about the process of telling, such as what the reteller knows about the listeners, also shape the story in the retelling. All three processes bring focus to the retold story. I retell what makes sense to me (understanding). If I forget parts of the story, the rest will be moved to the center (memory). And if I know to whom I am talking and why, I will adjust the story (purpose). My stories change if I tell them to my kids instead of a group of people at the bar.

Combined, the retelling shows how the reteller changes, simplifies, and embellishes all aspects of a story based on what she remembers and how she thinks the details fit the meaning she has grasped or created. For example, if the reteller thinks that a suspect in a legal case is guilty, she may make this clear in her account without even perceiving her adjustments as a change. The retelling then may just show what she thinks to be the mere facts of the story. Her suspicion could become the plot of the episode and thus be corroborated by her story. The doubts that she herself might still have no longer reach the next recipients of the report. In this way, suspects become culprits. It is also possible, however, that the retellers are precisely aware of the uncertainty and then emphasize it exactly the other way around, transforming the story of a concrete suspicion into that of a vague suspicion that no longer allows any conclusion. In the case of such a legal investigation, it seems, on the surface, to be about establishing facts to which the narration is subjected. But what's actually at stake here is something else. It's about the construction of a story in the interplay of details, causality, and emotional evaluations, as well as the fixation of feelings, such as suspicion. This chapter will focus on a precise exploration of this process of retelling.

Retelling is influenced by what we think the purpose of a story is and how we should adjust our story to bring about this purpose. This includes whether we think we need to entertain or inform others.[2] It also includes how we think others can or should understand our narrative. In a sense, we have a "theory" of retelling. This can be quite complicated. If you want to create an effect of surprise in your audience, you have to withhold certain information first, even if it's naturally in front of your own eyes.[3] You may even need to misdirect the audience, like a magician, to imagine something that will not occur. This means that the narrator must hide away his own overarching knowledge, so beautifully called *hindsight knowledge*, and pretend to not know, perhaps even bringing himself to actually forget, while still preserving it as sovereign knowledge of the story's design.

There are different ways to analyze retellings. One can look at individual retellings and learn how an individual processes a story. But one can also compare hundreds, thousands, and tens of thousands of retellings to see the tendencies that guide the changes. The guiding methodological question is always what remains of a narrative in the process of retelling and what has been changed. That is, how did a first story became a second? Retellings are thus a good means of gaining insight into the narratives in people's minds. Retellings reveal the tendencies of narrative people. Culture, in turn, is in a sense composed of the sum of all retellings as well as the transmission of knowledge and institutions over time.[4] Retelling lends itself to scientific inquiry in a wide range of fields, such as folklore, religious studies, history, political studies, legal studies, and psychology, as comparing different versions gives scholars an empirical basis of inquiry. However, retelling is at the same time a natural process that occurs over and over again in numerous situations. In the case of fairy tales and myths, retelling is eminently culture-forming. Building on this insight, eighteenth-century thinkers such as Johann Gottfried Herder and then the Brothers Grimm founded comparative cultural studies. Their belief was that specific cultures have a gravitational center that somehow informs the trends of how language, stories, and cultural artifacts are produced and transformed. A good hundred years later, Émile Durkheim formulated a general theory of culture based on ritualization, which of course is a form of stabilized repetition and retelling. And the early thinkers of psychoanalysis theorized repetition in retelling as a productive cultural technique (as in, for example, Sigmund Freud, *Totem and Taboo*). C. G. Jung's theory of archetypes also belongs here, since these archetypes are distilled only through multiple formations and retellings.[5]

More simply, and without following the theoretical assumptions of the early thinkers of psychoanalysis, the moment of repeatability can be understood as schema formation. In retelling, simpler and more concise forms usually prevail, and have stronger sticking power.[6] But the question here is, also, which schemas? And what exactly sticks? What motivates retellers to pass on a story, and what exactly functions as the anchor to which they attach the new story to the old and at the same time link it to new inventions?

This chapter will focus initially on three large-scale studies of retellings. Two of these are scientific investigations, experiments based on retelling games. Academics speak more technically of the process of serial reproduction. One person tells a story or reproduces an artifact to another, who in turn retells it to a third. In this way, many chains of retellings are created. Researchers can then compare all of the sequences and chains and determine which information was apparently most important to the retellers. Our third study of retelling is the collection of orally transmitted folk tales compiled by the Brothers Grimm, for these tales were passed down through generations and thus reproduced serially. In all three of our studies, the focus is generally on social information. Not surprisingly, social information tends to be passed down along the retelling chains preferentially.[7] But this does not tell us what form narratives take in the transmission of social information. As we will see, these large-scale investigations each reveal different tendencies of processing, remembering, and passing on stories, but they nevertheless will allow us to articulate a first theory of the narrative brain. Finally, we will compare these large-scale human retellings to the retellings generated by the artificial intelligence of a chatbot, ChatGPT.

Frederic Bartlett and Causality

A century ago, the psychologist Frederic Bartlett opened his laboratory at Cambridge University. He later described the scene as follows: "On a brilliant afternoon in May 1913 the present Laboratory of Experimental Psychology in the University of Cambridge was formally opened." But instead of enjoying the sun, he led the visitors into a darkened chamber. He wanted to trace the simple shapes that characterize our visual perception and also our narrative memory. He probably had in mind something like the system of elements in chemistry, with which all actual substances can be described as compounds of a few basic elements. There he sat, "in a darkened room exposing geometrical forms, pictures and various optical illusions to the brief examination of a long string of visitors."[8] The visitors' task was to recall the shapes. The hope was that the best-remembered shapes would constitute the basic forms of our thinking. In a reversal of Plato's Allegory of the Cave, then, it was precisely the impoverishment of sensory impressions in the darkened chamber that was to result in the true appearances and schemas remaining.

But it was not that simple. Simple shapes can be very complex. It took almost twenty years for Bartlett to summarize the results in book form. And what he presented there was no longer the doctrine of simple forms and elements, but rather a method of detecting stability in all forms of phenomena. This method consisted of asking visitors to recall a shape not just once but repeatedly over long periods of time, so that in the end the psychologist had a whole series of drawings. Similarly, a drawing could be passed around in chains from one person to the next. Each person who received it was supposed to look at it, then put it down, and redraw it with her own hand. The drawing then passed on to the next person. Consequently, Bartlett ended up with rows of drawings that he could compare.

Bartlett applied the same method to short narratives. A person was supposed to retell a story he had heard only once over and over again for long periods of time. Or a story was told to someone who would then retell it afterward, as in the telephone game (also called Chinese whisper), to another person, who would in turn retell it, and so on. In the end, in each of these cases, a series emerged that allows comparison and reveals whether the picture or story remains stable or undergoes changes. We know this activity as a party game under various names. Bartlett did not invent this psychological technique, but he popularized the method and called it serial reproduction. Today, this method of serial reproduction is used in various settings to simulate cultural evolution.[9]

Bartlett explained that this method was best suited for spotting the stable basic form in all phenomena and narratives. He called this form the "stereotypic form," from Greek *stereos*, meaning solid or stable. (Today's psychologists prefer the term *schema* rather than *stereotype*.)[10]

Fig. 2. A series of Bartlett's experiments in which different draftspersons had the task of repeating the immediately preceding drawing from memory (Frederic Bartlett, *Remembering: A Study in Experimental and Social Psychology*, 180)

His famous book *Remembering* (1932) provides a visual example of this process that can hardly be surpassed. A simple drawing of a barn owl stands at the beginning (fig. 2). The illustration follows an ancient Egyptian model. Barn owls do not actually have feathered ears, but the next draftsperson added small ear tips. Apparently, for that individual the owl pattern included ears. Soon a somewhat roundish, shaggy animal with ears, perhaps a tawny chick, emerges in the row. A subsequent draftsperson is then not very skilled with the pencil, and the little animal becomes a shaggy ball with ears. The next draftsperson is now faced with the task of interpreting this picture. Owl?

Mouse? One of them makes the call for cat, as the draftsperson before had already added a kind of tail. The shaggy animal with ears is smoothed out and becomes a cat. The wing line becomes a tail. And from then on there are no more problems. Cat is cat is cat. The series continues with a dozen highly similar cats. Whether the tail is right or left makes no more difference.

Bartlett suggests that the stereotypical animal with ears for his draftspersons in England is a cat and not an owl. This also shows that what is at stake here is not simple geometric shapes but the schemas in the mind that guide our expectations. For ancient Egyptians, the first image of the owl would likely have remained stable since it fit their cultural expectations. Barn owls were also the only animals that were not depicted in profile but frontally, and they had special prophetic significance. In this respect, reproduction is not a matter of creating some objective simplicity, but rather of linking the reproduction to those known forms a person considers meaningful in her environment. Researchers sometimes call this linking to known forms "priors," including prior beliefs, experiences, and expectations. In another series of painting abstract mask forms, for example, Bartlett shows how English participants transformed them into facial forms. These emerging faces are certainly more complicated in geometric terms, but at the same time they were much more commonplace for the participants in his experiments.

Bartlett also applied the same procedure of serial reproduction to short narratives in order to explore what the stereotypical form of narratives consists of. The differences between a picture and a narrative are, of course, interesting in many ways. In a narrative, the elements are delivered step by step in time and are not present simultaneously, as they are in a picture; expectation and suspense are essential components of reception; a wide variety of focalizations and identifications can occur in even the simplest story; linguistic turns of phrase allow for complex variations; and so on. Nevertheless, Bartlett made a strong claim about narrative reproduction in line with his findings of the owl-turned-cat series: narratives in serial reproduction have a clear tendency toward rationalization. Rationalization, he explains, consists in a high degree of "connexion of the parts," the absence of "incoherent" linkages, and the "simplicity" of processing so that the text could be "accepted by the observer."[11] The accent of this rationalization is accordingly on establishing causal linkages between the parts that are in line with prior beliefs people hold in regard to what is expected. The stereotypical form of narrative, Bartlett argues, is the causal narrative: why someone does something and why certain events occur. The cat of narratives is causality.

Indeed, causality seems like a strong candidate for stability and narrative memory. Who could have doubts about this? Well, I admit I had doubts. To examine Bartlett's claim more closely, it is important to look at the design of Bartlett's series of experiments.

As the starting point of his retellings and recollections, Bartlett chose a mythical story of the indigenous inhabitants of the Pacific Northwest. He justified his choice by the fact that this story and all its parts were certainly not known to his experiment participants in the United Kingdom. However, this choice had consequences. He chose as a starting point a narrative that likely seemed alien, incomprehensible, and thus not very rational to his participants. We should pause to question this rather unusual choice. Most narratives that people encounter in their everyday lives and conversations exhibit a lesser degree of strangeness, and this is precisely what makes them so appealing. In gossip stories from our friends, for example, we usually know the essential characters, or at least the narrator. That's why we want to hear the story. If the story is about Cynthia and Paula, it's better to know Cynthia and Paula, or to know they are important to the narrator. If we are asked to give a witness account of what we heard (or observed and will restate multiple times), we likewise know at least some aspects of the account and likely some of the people involved. Even in great literary and cinematic narratives, the protagonists are known and perhaps relevant to us in some sense, from Barbie to Oppenheimer and Harry Potter to Wonder Woman, and we may easily slip into their shoes or identify with them. In social media stories, we are likewise usually familiar with either the characters or the situation. "Yesterday at Whole Foods, I saw how . . ." That interests us because we either also shop at Whole Foods or have a negative affect about the wealthy who can afford to shop there. "Yesterday at the bazaar . . ." might likewise be familiar to some while evoking curiosity in others. The fairy tales we tell our children are usually also culturally familiar material. If, for example, a witch appears, we can immediately classify it and have concrete expectations, even if most of us probably know few real witches. We read books or watch series that give us plenty of time to get to know the protagonists. So we do not usually come across stories that are truly unfamiliar—situations that are alien to our culture or characters that are completely unknown. It would be equally odd to retell stories that are really none of our business. Thus, by choosing an unfamiliar mythic story from a foreign culture, Bartlett made a preliminary decision that created conditions applicable to only a tiny fraction of the narratives by which we are surrounded. And this choice is built on a story where most of his participants would have a hard time detecting its logic and causality.

The lack of intelligible causality or rationality of the source story for his experiment participants then drives the way they retell it. In Bartlett's story, ghosts appear and fight with some of the people. As Bartlett observed, one of his retellers changed the story by turning these ghosts into a tribe with the name "ghosts," thereby making the story more rational for the reteller. (Bartlett was amused by this fact and seems to have clearly understood what the true story is about. Unlike Bartlett, I have to confess, I am not confident that I understand the original story, and I feel like one of his participants. Certainly, Bartlett would have picked up on my German priors, as he did with his Indian participant.) Likewise, most retellers produce a more rational or causal story—for example, by naming when magic is used. But this is in reaction to a highly puzzling story. Whether Bartlett's results are transferable to narratives in general thus becomes highly questionable.

Bartlett is certainly correct in observing increasing rationalization in the retellings in this case of a story with low rationality for his largely British experimental participants. In the retellings, then, his three stated properties of rationalization can be noted: the higher "connexion of the parts," the absence of "incoherent" linkages, and the "ease" of comprehension (what is now called in scientific English *ease of processing*). In fact, Bartlett's theses are now the standard assumptions of similar studies, and his results have been successfully reproduced. No one doubts that causal sequences are important for narratives. Causality is part of most standard models of narratives today, such as the situation model (who does what to whom, why, when, and where).[12] Narratives usually provide clear rationales for why someone does something. Thus, we would also assume that it plays a role for narrative thinking and for retelling.

So, what could be problematic here? We must first say, in logical terms, that Bartlett's theses and approach involve a contradiction. On one hand, he chooses a mythical text, which, according to his own views, has itself already been retold many times and may therefore have already attained stability as a product of serial reproduction. On the other, it becomes apparent that even this seemingly stable text is subjected to radical changes when it is retold. The same is true of the stylized owl, which was probably perfected over millennia in ancient Egypt only to be transformed into a cat in Bartlett's laboratory by a few clumsy volunteers.

This contradiction is resolved when one considers that Bartlett here simulated a process of cultural appropriation and distortion—that is, assimilation. An alien and thus incomprehensible artifact of one culture is decomposed and reassembled in another culture. But these insights apply only to the process of cultural appropriation and assimilation and not to memory in general. It can be argued, of course, that any transmission from one person to another is always already an aspect of cultural appropriation, since every person thinks differently. This is true to a certain extent, but it does not take into account that most people in a cultural area and group tend to have similar ideas about things and are thus part of a culture.

In short, Bartlett's theses apply primarily to the special case of culturally alien texts and representations that are perceived by recipients as incoherent. These are then rationalized by the recipients. Rationalization in this case may indeed be an aspect of memory and retelling. But to what degree rationalization is effective in other cases remains unknown.

Bartlett himself did also work with everyday stories and he made some observations on the moods of narratives, but these studies have hardly been taken up in subsequent reception. Bartlett is known for his argument about increasing rationalization in terms of prior beliefs, which has become the basis of the standard model of narrative communication.[13]

In daily life, however, we are confronted with everyday stories and fiction that make sense to us. So let us begin our investigation with the very material with which Bartlett also began: myths, folk stories, and fairy tales. But instead of looking at culturally foreign fairy tales, we will look at texts that are reproduced again and again within one culture. As I will argue below, we are usually not primarily concerned with rationality and causality.

Vulnerability and the Brothers Grimm

In a well-known incident, two siblings were taken hostage but narrowly escaped from their captor. It is said that their wealthy female captor was a known cannibal. The circumstances of their escape from the mansion in which they were held show high levels of resilience and wit, as one of the children managed to lock the captor in her large kitchen's oven. In the absence of other relevant parties, the children were seen as the rightful heirs of this cannibal, who died in her own luxury oven. What do we learn from this tale? What are the rationalities or causalities of fairy tales? And more broadly, what made the Grimm fairy tales the most successful stories of the past few centuries?

Retelling and serial reproduction are a core element of our culture. In fact, it is no overstatement to suggest that culture would not exist without a recurring transmission process.[14] Over millennia, we have preserved our religions, myths, wisdom, morality, customs, laws, history, science, technology, family records, economic records, fiction, and so on in the form of oral transmission chains of stories. Looking at these story chains promises to give us major insights into what the stories preserve and what is the nature of a well-rounded, cognitively optimal story that we enjoy and can remember well.

One of the major places where serial reproduction occurs naturally is the transmission of myths and fairy tales, such as "Hansel and Gretel." Over long periods of time, such collective basic stories were passed on and spread in probably all cultures.[15] These tales should present the possibility of considering how stories were adapted to their culture and how they were optimized to fit our cognitive apparatus to process and retell stories.

Folk tales have a significance that can hardly be overestimated. They are often the only remnant left from past and lost cultures. In some cases, the oral transmission of these narratives has probably taken place from generation to generation over thousands of years. They convey religion, morals, a code of behavioral rules, and preparation for numerous situations. They also advertise good storytellers, who can pose as the heroes of the story, offer an identity to the group, and so on. But in the following we will, at least at first, be concerned not with the possible functions of folk narratives but rather with what is stabilized in the retelling. Is this also about rationalization, as Bartlett argues?

The most famous folk stories are the fairy tales collected around 1800 by Jakob and Wilhelm Grimm, who gave us Hansel and Gretel. These tales are not just relevant for those who are culturally German; they are also among the first printed folk tales globally and they have had worldwide influence with Disney and contemporary fan fiction. In fact, before the Grimms and several of their contemporaries, folk tales had not been seen as worthy of print and had not been collected. The first ever printed folk tale was published in 1777, only slightly before the Grimms, in an autobiography written by Heinrich Jung-Stilling (and edited by Goethe), a fairy tale that the Grimms then copied and included in their collection.[16] There are, of course, a couple of other story collections as interesting for us as the Grimms' tales, including the texts from the Vedic tradition in India and various biblical texts. (In the case of biblical texts, canonization hindered free-flowing adaption, although they are also highly instructive.)[17]

Given this earlier neglect of folk fairy tales, we lack earlier folk versions for most tales that we could use as a basis for comparison. However, for some other stories, we do have earlier versions that were penned by famous authors, and we will look at one of those examples later. Another approach we will take is to examine what can be learned from the Grimms' collection in and of itself and ask what these fairy tales actually conveyed and stabilized. In order to do this, we must necessarily also take a look at the epoch in which they were produced and collected. For that, we have to backtrack a little.

Children as Heroes

If Martians got a copy of the Brothers Grimm storybook in their tentacles, what would they think? Probably not much, because they would immediately adapt the stories to their perceptions and culture, similar to Bartlett's Britons who adapted the stories by Native Americans into their cultural world. We get a little further if we make historical comparisons. It is sometimes naïvely assumed that the Brothers Grimm simply wrote down stories that had been circulating orally for centuries. This is how the brothers themselves liked to portray it. However, as we shall see, the fairy tales represent a new narrative, that is, a new basic form that differed from preceding narratives.[18] This new form may have emerged relatively quickly and may also be the product of several storytellers and the editing practice of the two collecting brothers. In this respect, in describing the form of the Grimms' fairy tales, it is not helpful to talk about a preceding tradition. Instead, we must look at what distinguishes the narratives of the Grimm stories, which have been sensationally successful over the past two centuries, from other story types. The Grimm fairy tales still put every other international bestseller to shame. Why?

If we make a comparison between the prose of around 1700 and the fairy tales collected a good hundred years later, we quickly notice a change in tone. Around 1700, picaresque novels modeled on *Don Quixote* were popular stories. The most successful of these novels in the German sphere was *The Adventures of Simplicius Simplicissimus* (1668 or 1669) by Hans Jacob Christoph von Grimmelshausen, which was followed by a sequel novel with a female hero, *Trutz Simplex*, sometimes referred to as *Courage, the Adventuress* or *Mother Courage* in English. Both certainly contain a number of fairy-tale elements, including mythical creatures, transformations, and many near miraculous rescues. But something is also different. The heroes of the stories remain emotional and moral stand-up guys: great and terrible things happen to them. They survive when everyone around them dies, the heroine of *Trutz* gets raped, they cry or rejoice, and then things go on. Their fate often changes from one extreme to the other, and they usually quickly find their feet in the new situation, from galley slave to bon vivant, from hermit to warrior (in fact, Courage is the greater warrior of the two, as her tale is in general the more accentuated and includes more active choices). As we understand people now, we would expect someone to break under the pressure of the back and forth. But not Simplicissimus and not Courage. They get up again and again, flee, or just move on as if nothing has just happened. Whatever happens to them has almost no effect on the next section. Knight, robber, courtesan, nobleman, fool, and prisoner: fortune goes back and forth. It is as if nothing sticks to them except the good doctrines of the hermit, educator of Simplicissimus, and Courage's desire for revenge.

All of this is different in fairy tales. Of course, the heroes in fairy tales are not modern people in our sense either. We learn little or nothing of their inner experience. But the dangers they face threaten to break them. The heroes of fairy tales are vulnerable, and that is their central quality. That is why they are children and animals who appear weak in the face of the dangers that confront them and threaten to swallow them up. Fairy tales are built around the fragility of their heroes. The heroes are exposed to the world and are usually alone or in a small group of similarly fragile beings when the danger appears. The end consists of rescue from the danger, whether by their own efforts or by a deus ex machina like the friendly hunter in "Little Red Riding Hood."

Of course, heroes like Simplicissimus and Courage also find themselves exposed to multiple dangers. That is the nature of adventure novels, picaresque novels, and epics. But in these, the focus is not on the heroes' fragility but rather on their luck at avoiding danger or their ability to pull themselves up again afterward. This is not a subtle distinction: in the case of fairy tales, the focus centers on the inner nature of the living being; in the case of the picaresque novel, the focus is the next action. One can also put it this way: in order to make the Grimm fairy tale possible, the (inner) mutability of man had to be discovered. Simplicissimus and Courage remain the same in their inner being in every role. The fairy tale, in contrast, is built on the possibility that someone could become someone else from the ground up.

Discovery of Changeability

The inner changeability of a human being is a relatively recent discovery in the Western cultural sphere. In the eighteenth century, the Enlightenment placed great weight on the revaluation of man as a being that could be educated, formed, and thus enlightened. For this to succeed, a person needed to be recognized as changeable. This new way of looking at human nature was no small feat, and it reveals the enemy of the Enlightenment—namely, the idea of unchangeable qualities bestowed on a being at birth, qualities such as aristocracy and nobility. Because a person was proclaimed or discovered to be changeable, notions like inherited nobility became suspect. With this suspicion, the unequal starting conditions of people at birth became fundamentally problematic. With the idea of the changeability of man, the project of the Enlightenment became possible and at the same time de rigueur.

New ideas need strong images and narratives to gain acceptance. Even the mediators of the ideas must themselves first become clear about their idea, because new ideas are first and foremost new, unknown and misunderstood. Accordingly, a number of different metaphors are brought to bear on the changeability and mutability of human beings. One of the most prominent metaphors comes from biology: the "seed" of the "educational drive," or *Bildungstrieb*, which was brought into the field by the naturalist Johann Friedrich Blumenbach in 1789. The pedagogue Johann Bernhard Basedow had previously articulated this idea as follows: "When a child learns to read, write, and memorize in what belongs to religion, and is often chastised for carelessness or distraction, the seed of a future reluctance to such words and phrases is sown in the mind. This seed grows up."[19] In this line of thought, the educator can unintentionally produce seeds that have a long aftereffect. In Basedow's example, the seed is one of discomfort that will grow a vine that will remain intimately connected with the idea of religion. This seed is not present at birth but acquired—for good or for ill—through the process of education.

This non-intentional, unobserved, and non-conscious aftereffect is also the subject of another metaphor: imprinting. This appears especially in child-centered pedagogy, following John Locke's metaphor of the tabula rasa, Rousseau's *Émile*, and the writings of key figures of the German Enlightenment such as Basedow, Dietrich Tiedemann, Christian Gotthilf Salzmann, Joachim Heinrich Campe, Johann Heinrich Pestalozzi, and their companions.[20] In 1787, Campe formulated the concept this way:

> If you, good mothers, imagine the young soul of your new-born child as an extraordinarily soft lump of wax that willingly accepts even the slightest impression, then you are . . . on the way to forming a fairly correct image of it. . . . the soul distinguishes itself from all other known things in the world primarily by the fact that the impressions, which it has accepted once, can be changed by other following impressions, but cannot be completely smoothed out or destroyed by anything in the world. . . . Suppose, therefore, that there are certain colors of infinite variety, which, once applied, could not be completely erased again by anything in the world, or scratched out, or covered over in such a way that something of them would not shimmer out again and again. Imagine furthermore that all sensual objects in the world . . . would be equipped with such a brush, dipped in such colors, in order to paint everything, which comes close enough to them, without interruption. Finally, place a white board in the middle of these objects equipped with brushes, and observe what would come out of it? . . . Every sensual object that is close enough to your child to be able to make an impression on its sensory organs awakens an image, an idea in its soul. This image is temporary, and is immediately displaced by other images and ideas from the child's dark consciousness; but do not think that it has been completely lost for the little soul, or that it has remained without any consequences for it.[21]

This vision is the nightmare of every parent. Rather than drawing the picture of a strong and autonomous human being, Campe portrays a person characterized by affectability and changeability. Any fleeting impression will stay and stick, including any impression that eludes the guard of the parents. Today's parents would be thinking about the zillions of images of violence kids are exposed to on the internet.

The process can be summarized as follows:

1. Identity is acquired and not simply given.
2. This acquisition of identity happens in contact with the external world.
3. Contact with the external world happens by means of observation and perception. These are more passive than active (the objects paint the soul).
4. Because perception is the driving force, identity is continually changing. The resulting figure that Campe sets as the outcome of education also continually changes its form and surface, and it can be destroyed by a single blow.
5. Identity is unique; no two individuals are the same.
6. Identity is cumulative; nothing is lost even if it is covered by a new layer.[22]

In this lump-of-wax pedagogy, identity is gained only through contact with the outside world. Without this contact, the individual would be nothing but formless wax. The innate blood-and-soil identity of the nobility is ridiculed here in passing as nothing but a lump. The new form of an acquired identity allows for perfectibility but at the same time contains a great danger. Every impression of the external world has the potential to destroy the individual. Too big a blow and a dent will remain for eternity. Certainly, to stick with this image, whitewashing is possible. But the images underneath will not disappear completely and may eventually shine through. The stain of blood on the key in "Bluebeard" will always return.

It is thus appropriate to speak here of the description of *proto-traumatic imprints*. Although the repetition compulsion as a structure of the return of the suppressed is added only somewhat later, in German Romanticism around 1800, the structure of the violent marking from the outside prepares the traces that give characters their identity and impose patterns of action on them. From the wax pedagogy, there is a direct line of development that leads via Karl Philipp Moritz and E. T. A. Hoffmann to Friedrich Eduard Beneke's idea of the "trace" or "groove" and Sigmund Freud's "trauma." The past is thereby increasingly interpreted as a prison of the individual. Nicole Sütterlin has shown how pronounced the trauma narrative already was in German Romanticism around 1800 and how many protagonists of the novels and novellas suffer from trauma and are narratively explained by it.[23]

It should be briefly mentioned here that the trauma narrative has become one of the most important of the past hundred years. There is hardly a Hollywood film that does not locate a character's truth in his or her past. In the detective novel, the nature of the perpetrator is frequently explained in terms of the perpetrator's early influences and imprints. As people get to know each other better, it has become part of the ritual to tell each other about one's youth. Nothing explains so much about other people today as their traumas. Trauma is an important topic for narration research, because in trauma past events are used to make sense of the present state and future. That is, trauma stories reveal that narrative matters: the story of the past, if well understood, is also the explanation of the present world. This holds for clinical levels of post-traumatic stress disorder but also for "trauma" in the everyday sense: once we know that Severus Snape is being haunted by his memories of Harry Potter's mother, we understand his mean attitude better. In *The Silence of the Lambs* (1991) Clarice Starling aims in her present actions as an FBI agent to undo her traumatic memories; in the movie, the revelation of her past trauma to Hannibal Lecter leads to her receiving the clue to solve the case.

The idea of trauma (born from changeability) interconnects psychology and storytelling by making people more predictable. Indeed, empirical psychology finds one of its historical sources in the works of Karl Philipp Moritz, who started the *Magazine for Empirical Psychology* (Magazin zur Erfahrungsseelenkunde), wrote one of the first bildungsromans, and also wrote about mythology. Once the trauma of a person is known or posited, his or her motivations seem to be clear. This brings us back to the question of what was stabilized in the repeated narratives of the Grimms' fairy tales.

What Does It Mean to Be Vulnerable?

For a number of the fairy tales collected by the Brothers Grimm, there are earlier versions created by named authors, such as Giambattista Basile and Charles Perrault. These authors might well have taken up oral stories available to them, and then their own spiced-up versions might have had their impact on oral traditions as well until they were then taken up by the Grimms via oral transmission. What changes can be identified in a process such as this?

We should not be surprised that the oral versions have omitted many specific details. And, to be sure, omissions can change the essence of the story. Take, for example, the ending of King Bluebeard, one of the Perrault fairy tales that the Grimms included, in its orally transmitted form, in their first edition of the fairy-tale collection of 1812 but omitted in later editions.[24] In this fairy tale, a young woman, recently wed to King Bluebeard, is entrusted with a key with the instruction that she must never open the chamber in the castle to which this key belongs. With the announcement of the prohibition, what will happen is clearly signaled: she does the forbidden thing, finds the bodies of King Bluebeard's former wives in the chamber, and narrowly escapes the same fate by killing Bluebeard with the help of her brothers. Perrault ends the tale as follows:

> It turned out that Bluebeard had no heirs, so that his wife became the mistress of all his riches. She used some to marry her sister Anne to a young gentleman who had loved her for years; some she used to buy captains' commissions for her two brothers; and the remainder, to marry herself to a man of true worth, with whom she forgot all about the bad time she had had with Bluebeard.[25]

The narrative ending aims at a precise balancing in which everyone gets what he deserves: Bluebeard is killed, and the others receive carefully weighed rewards for their help or for the evil suffered. This is comeuppance as a lawyer might like to see it. The compensation succeeds: in the case of the heroine, it is said that she now "forgot" the evil days. Perrault thus assumed that what has happened can be undone, which again emphasizes the structure of a balancing equilibrium. But that is not enough because Perrault follows the tale with the moral of the story, and this in two forms:

> The Moral of This Tale
> Curiosity's all very well in its way,
> But satisfy it and you risk much remorse,
> Examples of which can be seen every day.
> The feminine sex will deny it, of course,
> But the pleasure you wanted, once taken, is lost,
> And the knowledge you looked for is not worth the cost.

> Another Moral
> People with sense who use their eyes,
> Study the world and know its ways,
> Will not take long to realize
> That this is a tale of bygone days,
> And what it tells is now untrue:
> Whether his beard be black or blue,
> The modern husband does not ask
> His wife to undertake a task
> Impossible for her to do,
> And even when dissatisfied,
> With her he's quiet as a mouse.
> It isn't easy to decide
> Which is the master in the house.[26]

The first moral simply blames the protagonist for what happened: her curiosity almost brought her down. Here, then, is classic moralization. The female affect of curiosity caused the catastrophe. The woman's naïve curiosity "is not worth the cost." The second moral, in contrast, with its obvious irony, offers a contradictory interpretation. First, it denies she was really at fault, for it is now implied that the man asks "*a task / Impossible for her to do.*" Curiosity may have been her impetus, but this is not to be blamed on her but rather on the man's impossible assignment. By giving her this order, he created her burning curiosity and thus actually caused the transgression of the taboo. His misdeed here is not so much the murder but rather generating her curiosity, thus casting her in a feminine role. The male affect of jealousy is now to blame, as the modern man is not "master in the house" any longer. Perrault uses the story for a sideswipe at his epoch, in which the hierarchies of man and woman have softened and women have risen in station. The fact that in both morals women come off badly, being either naïve or unnaturally masculine, should be also be noted here.

Perrault's ending with a doubled moral thus accomplishes a quadruple purpose:

1. It sanctions and rewards all survivors.
2. It balances out what has happened and suggests it can now be forgotten.
3. It invites close consideration of guilt based on motivations and affects.
4. It creates ironic distance with a comparison to the present time.

In contrast, the end in the Grimms' version consists of a single sentence after the killing of the culprit: "Then he was hanged in the blood chamber with the other wives he had killed, but the brothers took their dearest sister home with them, and all Bluebeard's riches belonged to her."

The end is short, but not painless. The evil is averted, punishment befalls the culprit, he falls into his own trap, and he is executed next to the victims. Comeuppance here becomes poetic justice. All reward falls to the woman. No further mention is made as to whether the brothers are also rewarded, and, above all, no precise evaluation of guilt is undertaken. The quick ending throws only a faint veil over the horror, which is also named in the last sentence with its "blood chamber." Without assignation of guilt, morality, and ironic distancing, there remains only the bare act of being saved, but there is no forgetting. The woman is rewarded by means of material inheritance.

This shortened ending sheds light on the fairy tale as a whole. When the narrative arc ends in the mere state of being saved, the emphasis falls on the moment of threat and danger. The Grimms' fairy tales, unlike Perrault's, do not invite the specific consideration of guilt. It seems that here—also unlike Bartlett's rationalization—it is not so much about complex chains of causation. Instead, the focus is on the threat of imminent danger. Danger and threat constitute the situation of the story. Accordingly, the young woman is saved, simply because she is in acute and horrible danger. To put it simply: Grimm fairy tales are all about their heroes' vulnerability. All other elements of the story are centered on this vulnerability. Other aspects, such as the moral, the pondering of guilt, and the names of the protagonists are omitted. The elements of Perrault's story as well as the templates of other fairy tales are adapted accordingly. The seducible noblewoman in Perrault becomes a vulnerable girl in Grimm. All of the narrative events in the Grimm fairy tales, including the characters' identities and their world-views, are oriented toward this possibility of wounding.

What does it mean to be vulnerable?

1. First of all, vulnerability implies that one is affectable and thus changeable in the broadest sense. Someone who is vulnerable is not clearly fixed and has not reached a stable identity or final state. This person's identity is still in flux and can be affected. Or, expressed differently, the vulnerable being remains accessible to others (*affectability, unfinished identity*).

2. At the same time, vulnerability implies a threat. The vulnerable person is constantly in danger (*destructive potential*).

3. Vulnerability thus implies a reactivity to injuries: the vulnerable being has some basic awareness of injury and so develops preferences to avoid injury and strategies to escape the source of danger. The state of vulnerability, one may conclude, suggests to the individual the goals of avoidance and of healing after actual wounding (*seeking protection, immunization*).

4. Whoever is vulnerable, one may further assume, learns to evaluate the environment for its potential danger. The individual registers a stimulus coming from outside and recognizes its influence (*value standard, evaluative learning*).

5. Paradoxically, this structure can imply a degree of danger-seeking as well, as the vulnerable being is probing what causes injury to increase awareness. There may be further implications for traumatic restaging of the scene of wounding (*traumatic return*).

Vulnerability can be told, and the Grimm fairy tales tell vulnerability as a narrative pattern:

1. A being is marked as weak; it may be, for example, a child or perhaps an animal.
2. The being is isolated from others in many respects. For example, the being is an orphan and has an evil stepmother. There is either no social network to protect it or one that only consists of weak beings. Then the being is thrown into the world alone (or with a similarly weak companion) or it escapes; it may also be kidnapped or abandoned in the wild (isolation, abandonment).[27]
3. The external world is strange and threatening; it is like an opaque forest where many dangers can be suspected (threatening environment).
4. In most cases, the danger comes to a head. A character personifies the evil, perhaps becoming a wolf or a witch. This hostile character is not simply an opponent or a human rival, but is instead overpowering and dreadful (personified danger, mythical disaster).
5. Now the child is symbolically or physically wounded—for example, by transformation, injury, or murder (wounding).
6. In keeping with the stylized and often emphasized danger or wounding committed by some overpowering mystical being, the rescue then also occurs in radical and emphasized form, perhaps through the arrival of a deus ex machina. The source of the danger is destroyed, or at least the child is permanently saved (rescue, or magical revival).

The folk tale tells this vulnerability narrative in many variants and optimizes it. The protagonists of fairy tales are usually young people, nameless boys and girls, princes and princesses, or animals. They are exposed, isolated, deprived of their parents, or otherwise defenseless. Vulnerability practically gives birth to danger, narratively produces the heightening in wounding, and enables the narrative arc. A terrible wolf can peer from the face of any grandmother; Bluebeard's blood chamber can open behind any door. Another aspect of vulnerability here is seductiveness: the children, of course, want to nibble on the gingerbread house. The locked chamber is all too tempting, as is the forbidden forest. The children are vulnerable to seduction, and they in turn are seductive since people are attracted to the cute and weak. Here we come back to the central thesis. In the Grimms' fairy tale, the condition of vulnerability becomes the condition of narrativity: *because vulnerability exists, narration can take place.* Vulnerability and narration are mutually dependent. And, as readers may have noticed, I also suspect connections to the structure of modern subjectivity in the narrativity of vulnerability.

The Structure of Cunning

So far, it seems as if the characters in the Grimms' fairy tales are purely passive pawns of events. But this is not the case; they themselves are also agents, acting and making decisions. In other words, they are also heroes.

This concerns, first of all, the fact that defenselessness can also be the result of a more or less self-inflicted transgression of the taboo or a miscarriage. In "Iron John," for example, the boy gives the wild man the key to his cage; in "Bluebeard," the young bride opens the forbidden chamber; and in "Jorinde and Joringel," the first folk tale ever printed, the children penetrate the forbidden circle of the witch.[28]

But even more important, the children of fairy tales can also find their own way out of danger. They themselves outwit the wicked witch and throw her into the oven, trick the devil, defeat seven great giants using nothing but their cunning, find the name of the enigmatic dwarf, or dream up the solution as in "Jorinde and Joringel." Even a tomcat can ascend the throne as adviser to the prince, the miller's daughter can become queen, and the children can triumph as heroes.

The sequence in this narrative follows the tradition of the cunning hero story, which includes epics like Homer's *Odyssey* and picaresque tales such as those about Robin Hood or the German folk hero Till Eulenspiegel. Vivasvan Soni has called this a form of test narrative: the heroes are challenged to find a solution. For this they may have to endure suffering (some dose of vulnerability), but in the end they can emerge victorious from the test.[29] The cleverest ones win because they outwit the overpowering enemies, often by turning their own weapons against them. The end rewards these cunning heroes with gold, wealth, wonderful gifts, kingdoms, handsome princes or beautiful princesses, or a triumphant return home.

What is striking is that both narratives—that is, the narrative of vulnerability and the test narrative—overlap in the Grimm fairy tale. Hansel and Gretel show the vulnerability of abandoned children *and* their cunning in fighting the witch. Joringel is enchanted by the witch *and* dreams the solution. Puss in Boots first falls victim to his young master's poverty *and* then finds the tricks that gain them a kingdom. The Bremen Town Musicians are outcast, weak animals *and* in the end come to a life of prosperity through their cunning.

The tension between the two narratives is enormous, and yet both seem to enter into a simple union in the Grimms' fairy tales. I will suggest in the following that it is precisely this joining and overlapping of two narratives and lines of tradition that constitutes the formula of the Grimms' fairy tale. How do the clear passivity of vulnerability and the reward of the cunning or courageous come together?

The Double Fairy Tale: The Reward of Vulnerability

My overall claim here is that fairy tales discover vulnerability as a virtue and a powerful narrative tool. This is what I would like to suggest: the two central narratives in the fairy tale are the vulnerability narrative and the test narrative. The first leads from radical vulnerability to a state of danger or even actual wounding, until rescue emerges at the point of maximum danger or misery. The second narrative arc exposes the hero to a dangerous test that he can pass or fail. The danger comes and is overwhelming, but it can be defeated by deception, cunning, and the young hero's daring. These two narratives join at the end point, namely, the rescue from and triumph over danger. The effect of joining the ends is that both starting points—that is, vulnerability and cunning—are blessed by the qualities of the other: vulnerability appears as a winning strategy and cunning seems morally laudable. Both vulnerability and cunning appear as pseudo-moral qualities that lead to and deserve to lead to a happy end.

There are many fairy tales in which both narratives form such a combination. Vulnerable Hansel and Gretel are endangered and prove to be cunning, or at least witty. Even in this combination, the tensions remain; the heroes are described as mostly victims who are at the same time cunning actors. In the joining of two narratives, the two sides of Freytag's pyramid that we described above come together when the heroes passively suffer on one hand and actively save themselves on the other. (There is a somewhat similar tension in storylines that have an underdog as hero, but in that case the narrative is fueled by the rivalry with the top dog.)

It is particularly important that both narratives are resolved together at the end. The vulnerable child is rescued (end of vulnerability) and the enemy force is defeated through courage and cunning (victory of cunning). In many cases, the rescue is a return home ("Frau Holle," "Hansel and Gretel," and others) or marriage, the establishment of a new home, and the like. In any case, it brings a reward. Gold, money, and a royal wedding are standard. This reward is the sanction concluding the fairy tale. Because the sanction at the end regularly concludes *both* narrative strands at once, it rewards not only the cunning means of passing the test but also the state of vulnerability as such. Vulnerability becomes a value and a strategy that can lead to victory. This is remarkable, inasmuch as the vulnerable person does not always deserve a reward. The vulnerable person has usually not done a service to anyone (at least through vulnerability itself) and is not morally superior.

Here we come to a surprising conclusion: in the double structure at the end, the fairy tale thus sanctions more than just the passing of a test. The vulnerability itself is positively valued as if it passed the test. The gains from the cunning and the compensation for the wounding merge into one reward in the fairy tale. The structure is that of a chiasm. Usually, the passing of a test (cunning) deserves reward and triumph, and vulnerability deserves care and protection. In the fairy tale, the outcomes merge and cross over: the passing of a test becomes morally superior and the vulnerability becomes triumphant. The sanction elevates the state of vulnerability and cunning jointly into a moral realm of triumph: because there is a reward at the end, vulnerability and cunning must have been morally good. The vulner-ability-cunning plot acquires the status of an as-if morality through the positive sanction at the end. Being vulnerable does not just mean that one deserves protection, but becomes a moral action associated with triumph. This is the grand narrative of the fairy tale that has had transformative power in the past centuries.

The fairy tale rewards being exposed and the state of affectability itself and elevates them to a value. The ideology of the Enlighten-ment, that the good man is the changeable man, finds its narrative confirmation in the fairy tale. The fairy tale is thus brought into the debate as evidence in the dispute about other types of identity, such as that of the unchangeable, birth-given nobility. The fairy tale exhib-its vulnerability and shows, through the positive feeling of the recipi-ent and the victorious outcome, that vulnerability is morally, socially, and emotionally superior.

Vulnerability is, at its core, the possibility to change and be af-fected. The vulnerable person experiences pain and suffers, and this means he or she is receptive. In that sense, vulnerability is also the core structure of being able to receive or perceive emotions. By re-warding vulnerability, fairy tales turn receptivity and emotionality into the key quality of humankind. Even better, this quality also becomes a winning strategy: the audience learns that those who reveal their vulnerability will win. Today, claiming one's vulnerability can be used as an act of empowerment.

Vulnerability as Ideology and Weapon

Vulnerability is positively sanctioned and elevated as a value in the fairy tale. We have come to find it emotionally moving to think of vulnerability, and this emotional response is likely fueled by our fa-miliarity with the fairy-tale plot that leads to a happy ending. It must be emphasized again that vulnerability in itself is not a moral value that helps others. In the fairy tale, the heroes are rewarded for merely displaying vulnerability. The most vulnerable wins. Vulnerability thus reverses the values of the preceding epoch. Stoic demeanor, imper-turbability, steadfastness, and thus invulnerability were held in the highest esteem until about 1750, as dramas such as *The Dying Cato* by Johann Christoph Gottsched show. Around 1800, on the other hand, traumatic injury was discovered and to some degree also invented— that is, the recognition that some injuries continue to have an effect even after the physical wound has healed or even without such a wound.[30] This idea of trauma already guides Basedow's and Campe's pedagogy, which is concerned with the aftereffects of the imprints of childhood. In these accounts of the late Enlightenment and the Romantics, trauma is not an avoidable evil, but the condition of education and upbringing. Both this reevaluation of values around 1800 and the discovery or invention of trauma will be mentioned only in passing here, but they are a key offspring of the elevation of vulnerability.

Vulnerability shakes the fabric of the narratively organized world. It is no longer the most resistant who wins, as in the tragedies of the seventeenth and eighteenth centuries, but the one who is attacked the most, in the double sense of being attacked and being vulnerable.[31] We could describe both the previous stoic regime and the new age of vulnerability as ideologies. And like any ideology, power lies in controlling narratives. An ideology of vulnerability exchanges the negative moral imperative that one must not hurt anyone with the positive twist that the vulnerable is intrinsically good; accordingly, it becomes desirable to be vulnerable and to show vulnerability. "Frau Holle" (or "Mother Hulda") can be seen as a training ground for this vulnerability. In this tale, two stepsisters are tested in regard to their empathetic vulnerability, but only one of them passes the test. She is willing to suffer for others, while the other one only eyes the reward at the end. The narrative unfolds vulnerability first into a means of sensitivity and this sensitivity into a value or goal. Means and ends are flipped around: sensitivity for others starts to be not just a means to an end but a goal and a reward.

The advancement of vulnerability certainly has been a central force of democracy. The narrative of vulnerability has likely drawn increased attention to suffering and oppression by connecting them with a sympathetic affective response. However, as a cultural technique and affect ideology, this narrative can be employed and utilized for various ends. Vulnerable people are sensitive people, but so are narcissists.

Those who style themselves as vulnerable people, even shedding a tear at the right moment, are already doing something seen as good. In short: vulnerability can become a weapon. Those who show themselves to be vulnerable win. Being stoic is out in the era of the fairy tale.

Without doubt, this narrative of vulnerability has played an important, often central role ever since, finding expression, for example, in psychoanalysis, modern pedagogy, nineteenth-century labor movements, twentieth-century feminism, and twenty-first-century movements such as Black Lives Matter. The notion of vulnerability in the discussion of recent decades also shows an apparent proximity to the emphasis on vulnerability in the Grimm fairy tales. The importance and valorization of vulnerability at present are certainly in no small part an achievement of feminism and the equal rights movements.[32] Also, #MeToo and Black Lives Matter clearly demonstrate that one's wounding, the trauma suffered, and systematic injustices are no longer taboo subjects that condemn victims to quiet solitude and shame, but now carry highly persuasive and argumentative power.

At the 2021 Olympics, for example, the amazing gymnast Simone Biles drew much applause for not competing in many events because she did not feel mentally strong enough. Similarly, the tennis player Naomi Osaka repeatedly reported her depression and anxiety, and this did not hurt her image either. What is interesting to us about this reappraisal of vulnerability today is that there is a marked difference between vulnerability in the fairy tales of the 1800s and the newly discovered vulnerability in recent years.

What is new today is that displaying vulnerability has become an accepted option. In public and political disputes, one can now declare oneself affected and reveal one's vulnerability. Those who claim vulnerability for themselves or others gain attention and bring into the open the hidden discriminations that only become clear when they are made public, precisely because someone declares himself to be affected. This is a very important corrective. More victims have dared to come forward than did so in the past, thereby doing a service to many others in signaling that violence can be exposed and prosecuted.

At the same time, however, it can also create the impression that someone is right simply because he or she claims to be affected and demands recognition. These declarations or complaints are a new technique that did not exist in fairy tales. The vulnerable and exposed children could not escape their dangers by declaring themselves vulnerable. Nevertheless, there is a similarity here in that a moral claim is made that does not derive directly from the ethical dimension of a behavior, but from the rhetoric of representing an affective situation. Vulnerability per se does not constitute evidence of being in the right, but this very impression can be created.

There is definitely a similarity here to the discourse of honor in the nineteenth century, with gender roles now partially reversed. Those who declared their honor violated then demanded a duel and satisfaction.[33] Of course, today's situation is different, because the injustice suffered and the trauma of the victims cannot be compared to the petty insults that could provoke a nineteenth-century gentleman to defend his honor. The similarity lies in the demanding action of declaring oneself injured. This opens up another form of behavior, enabled by the new grand narrative of the Grimm fairy tales.

Let's return to our question: what do the fairy tales reveal about our narrative thinking in general? First, the core of the fairy tale is not causality and rationalization in Bartlett's sense. The tales are often quite absurd in that regard, and how a particular rescue happens may not matter much. Instead, the Grimms' fairy tales show an optimization of vulnerability. At the core stands the affect of seeing the weak rewarded. The rescue of Cinderella, who will occupy us later, is exceptionally moving. Second, narration is about change and events. We are interested in those stories that cause an alteration in the protagonists. The protagonists emerge as different beings, transformed by the story events. Changeability, in this respect, is a central condition of narrativity. In vulnerability, this very condition of narrativity is embodied in the protagonists: narrative thinking is most successful when it includes the possibility of changeability-vulnerability of the protagonists. Narratives are about the possible imprints, that is, the dangers and positive possibilities of human beings and sentient beings to be affected by narratives. This logical circle stands at the core of narratives. In short, these fairy tales show and propagate the foundation of narrativity: presenting characters as open to change means that what causes the change matters, and that is the narrative itself.

Emotion Retelling at the Experimental Humanities Laboratory

So, what is the core of story retelling? What do we remember of a story? Bartlett suggests that causal chains are preferentially passed on in certain cases. The Grimm fairy tales show that causality is not an end in and of itself but can also be taken into the service of an affect ideology. What is the relation of causality (that is, the story why someone did something) and emotion (that is, how a story appears to us emotionally)?

Scientific studies on the retelling of emotions have been few.[34] Moreover, there are few studies of everyday stories with reliable data. There may be more studies of retelling in chains replicating Bartlett's findings with an exotic story than studies using everyday stories. I guess seeing a rare yellow bird seems more special than a thousand sparrows to many researchers. I am sure that I am guilty of this bias myself in many cases.

My collaborators and I suspected that emotions play a central role in narratives. Since we could not find studies that would help us, we conducted a series of studies with retellings using the serial-reproduction method. In doing so, we were concerned about statistically reliable statements, and we conducted experiments with many participants. In one series, about 12,000 participants were involved and generated almost 19,000 retellings that we could use. I should stress that every story was read and evaluated by three to six of our research participants (and we recruited many thousands). We also read most of the retellings.[35]

In these studies, we wanted to find out how the emotional dimensions of a story situation are passed on. Emotions, feelings, and affects, as well as moods, exist at many levels in a narrative. First, the characters have feelings, for which there can be many reasons. Then there are the situations in the narrative that prompt emotional appraisals—when a situation is judged to be sad, dangerous, or disgusting, for instance—whether or not the characters in the narrative have these feelings or affects. Finally, recipients have emotional reactions to the narrative; they may, for example, feel sorry for characters, share feelings empathetically, or develop an aversion to the narrative as a whole.[36]

We focused particularly on the emotional appraisals of the narrative situations. The reasoning behind our choice was that narratives are so appealing because they allow for co-experiencing, and this co-experiencing likely has a lot to do with the emotions of the key situation. As recipients of a narrative, we find ourselves transported into the situation of the story, we react to the situation, and, to some extent at any rate, we have precisely the experiences that one can have in such situations. This is where emotions come in. Accordingly, the emotions that emerge from the situations are a key part of the narratives. At least that's what we thought.

Our suspicions were quickly confirmed in one respect. Despite all the differences between the people, there was a great deal of agreement among our test participants about how happy, embarrassing, or sad the situations in the initial narratives were. Apparently, narratives are very good at helping us imagine how we would feel in a situation. And that is precisely what we were concerned about. If people associate specific situations in a narrative with similar emotional appraisals, then presumably that emotional appraisal plays a role in the narrative as a whole. Happy endings in stories are different from, say, narratives that end with disgusting situations.

Still, what role do these emotional appraisals of a narrative situation play in our memory and, in our case, in the transmission of the story? Is emotional appraisal a by-product of what happens, or might it even be at the center of our reconstruction of a story? Take the example of a company Christmas party. Many people are talking, the manager is handing out mulled wine, but the new department head is getting hopelessly drunk. This leads to the end of the party. Our retellings may focus on how embarrassing the behavior of the new boss was. The details—what exactly he did, whether he threw up in a vase or spilled wine on his employee's blouse, and where exactly it happened—are secondary.

We need to stress that retelling emotional appraisals arising from complex narrative situations is not simple: it requires arranging complete scenes. Anyone who wants to relay an emotional assessment must have developed a "theory" of the impact of the narrative. To be sure, our starting stories did not include words conveying emotions such as *embarrassing* or *sad*. Consequently, retellers could not simply copy a label of the emotion that fit the story. Instead, they had to generate the narrative situation of what people were doing and experiencing to bring about the affect or emotion. We felt that this would allow for a fair test to see whether the emotional appraisals of small narratives are passed on from one person to the next (and not whether words describing emotions would be selected more strongly).

Before reporting what happened, I have to get a bit more technical. We varied our procedures for our series of experiments to accommodate different scenarios. Experimental instructions always shape the results, so let me explain this briefly. In one study, we asked participants to tell us a sad or happy story in about twelve to fifteen sentences. We then had these stories rated by a variety of other participants in terms of the degree of sadness and happiness and other criteria, such as perceived causality. Other participants then retold the story for us in the serial-reproduction procedure. When we asked them to retell the story, we never pointed them to emotions, causality, or anything else, because we wanted to reduce so-called priming effects. We just asked participants to retell the story in "your own words." After all the stories were retold in three separate generations or iterations, we recruited different participants for evaluating these stories. We asked them how the events were causally linked, how happy or sad the situation was, and to what degree they could see the events in front of them.[37]

In another series of experiments, we ourselves generated a number of basic stories that contained situations that were charged with a range of distinct emotions or affects. For example, something embarrassing happens to a slightly shy person during a public speech. A new student is somewhat lonely on campus until some positive turn of events. A couple on a trip in Southeast Asia gets into a situation after their car breaks down on a dirt road and they are offered food by the friendly locals that is disgusting to them. A group of young people put themselves in danger at the top of an abandoned skyscraper. In these basic stories, we created a large range of variation in the emotional situation to produce a wide range of emotional intensities. In the case of the lonely student, for example, there was a positive event, but this could be the onset of spring with better weather (mild joy), an improvement in her study situation (medium joy), or bonding with a warm group of friends (strong joy). Again, we did not determine the intensity of joy or the emotions but asked a large number of participants to rate emotions and other factors of the stories. In this way, we produced three different baseline stories for five different emotions (disgust, joy, shame, risk/fear, sadness) and in turn provided five to seven variations for each from mild to strong with different endings, arriving at a total of one hundred variations. Subsequently, we gave these stories to others to retell them. Each starting story was given to thirty to fifty different people for retellings. We then gave these retellings to another set of retellers, and their retellings to another set of retellers. In this way, we produced thousands of parallel chains of retellings. This procedure seemed to us the best way to capture and compare differences in emotional intensity and to have conditions where retellings could be compared to other retellings.

It should be emphasized that all story ratings in our experiments were done by humans, in contrast to most studies today, which rely on automated or semi-automated sentiment ratings. Automated sentiment detection has certainly strongly improved in the last few years, but it uses procedures that are still not well suited for narratives. A common method of sentiment detection uses individual word scores according to a catalog and then simply adds them together. Contexts are necessarily ignored. Here, for example, is a sentence of only positive words: "The young woman's dog barked joyfully in the house as the young woman took her suitcase and went into the garden to travel to America for her new job." Humans understand that the dog will be abandoned, but automated word-by-word ratings will only view this sentence as positive. Some methods of sentiment detection include immediate contexts of words to gain in precision, but many of the subtle nuances that give meaning and emotions to a narrative situation are necessarily lost in the process.[38] The gold standard, so far, are human ratings. Narratives create contexts, and these are what narrative situations are all about. The question is whether and how exactly these are transmitted.

This question arises especially because any retelling produces numerous changes, as all the studies by Bartlett and the analyses of Grimm fairy tales presented here have shown. On one hand, narratives are great for recollection and transmission, as pointed out by many memory artists who make up mini-stories to recall incoherent facts. Moreover, narratives also motivate transmission. On the other hand, each reteller gives personal twists to the story that can change the subject of the narrative in any way. In this regard, many retellings may primarily appear as abridgments. In our studies, too, we recorded shortenings of about 30 percent from generation to generation, an effect often referred to as leveling. This makes more urgent the question of what is actually retained as each abridgment is, necessarily, also an interpretation.[39]

Our results came in. Our more than 12,000 participants had produced and analyzed close to 20,000 stories. Where Bartlett had observed increases in rationality and causality, we registered continuous declines from generation to generation in all three of Bartlett's aspects of rationality (a high degree of "connexion of the parts," the absence of "incoherent" linkages, and "simplicity" or ease of processing). The case was the same for story quality characteristics: only boringness increased. But what happened to the emotional ratings? At first glance, the results seemed to indicate a miss. When we compared the four generations of happy and sad stories from the original to the third retelling in terms of the degree of happiness or sadness, we could not see any clear changes. The sad third retellings stories were still sad to the same degree as the ratings of our original stories. The clearly happy stories had also retained approximately the same degrees of happiness. That is, in terms of the emotional ratings of happiness and sadness, not much had actually changed.

That didn't seem particularly interesting at first. When I lend my bike to a friend, I expect it to come back largely unchanged. But then we took a closer look at the stories and chains. Not only had they shrunk, and after three retellings were less than a third of their original length, they often showed a lot of creative freedom in their rendering. Here's an example. The original story, invented by one of our experiment's participants, reads as follows:

> I was sitting in my office when I noticed a moderately sized spider walking on the floor next to me. I picked up a notebook and small box and crouched down to scoop the spider into the box. I placed the notebook over the box to secure the spider inside and walked towards the door leading outside. Once I got outside, I removed the notebook and saw the spider sitting inside the box. I turned the box upside down and gently shook it to encourage the spider to go into the grass below. I looked at the ground and the spider was not there. I looked inside the box, and the spider was not there either. I was curious as to where the spider disappeared to. That is when I tilted the box to the side and noticed the spider rappelling from the box on his web! I smiled and then slowly lowered it to the ground and freed it in the grass.[40]

What should be emphasized in this story is the moment of aesthetic beauty when the spider is moved outside to the garden but then seems to disappear as if by magic: *"I tilted the box to the side and noticed the spider rappelling from the box on his web!"* Our experiment participants rated the story on average with a relatively high value of happiness (4.9 on a scale from 1 to 8). The well-told story was also rated as very coherent (5.8 on a scale of 1 to 8). Afterward, following three retellings (given by three different retellers), the story had shrunk to the following short version:

> I tried to direct a spider out of my office, giving it a chance to leave. It stuck around and I actually started to like it more and now we're friends.

The rescue-and-release story turned into a friendship story. In this case, the rating of happiness increased slightly to 5.6 on average, and coherence decreased only slightly to 5.0. Statistically, then, there is not much difference to be seen here at first on the basis of these data. But the reading shows how different the stories are. The entire garden episode is reduced to a metaphorical phrase: *"it stuck around."* While the original story saw the spider building a web, the reference to sticking now metaphorically refers to staying put. It is also particularly striking that the whole removal action has been erased. Instead, the new story simply says that man and spider are now friends. In what respect can friendship be considered a substitute for rescue? In regard to the level of positive sentiments. The good intention and action of rescue is rewarded with friendship. The abbreviated story here anticipates the positive outcome and leaves the reader with what was obviously most important: the alliance. Even in this reduced story, we see the minimal structure of Freytag's pyramid, discussed in Chapter 1. The active action (giving the spider the chance to escape) is followed by a passive experience of the human, as he is rewarded with friendship.

What is the pattern here? Let's look at a second story, invented by a different anonymous participant:

> A man named Fred had been called in to work on a weekend to help with an emergency. When he got to work, a client had been waiting and needed immediate help. Fred began to help the client, and asked where the client was going to in such a hurry. Fred learned that the man had a broken down car, and needed to pay for repairs quickly to get to his mother's funeral. Fred felt so bad for the guy so he asked the client if he would like to get a ride. The client was so happy, and Fred was proud and glad to help the man out. The client went online and then shared his story with other customers who were shocked by the generosity. The next business day, Fred went in to work and there was a large check waiting for him in his lobby, and he asked the receptionist where it came from, but the receptionist told him that it was an anonymous drop off.[41]

This story, with an average happiness rating of 5.6 and a coherence rating of 4.3, features a relatively complicated but logical sequence of events. A man named Fred helps a client and drives him to a funeral. The client posts a story about Fred's helpfulness, which in turn prompts an anonymous third party to write a check for Fred. In all, including the anonymous donor, the mother, and the receptionist, the short story mentions five people, which lowers coherence somewhat. After three retellings, the story looks like this:

> Fred helped a man who had to get to a funeral in a hurry. When he returned to work he found a large donation had been made.

This crude abbreviation simply drops the causal intermediate step—namely, that the man posts the event on social networks. Accordingly, the donation comes out of the blue. Instead of five named people, two remain, and it is unclear whether the man himself makes the donation or an unknown person. Happiness remains, but coherence no longer does (average rating of happiness is 4.8 and coherence 2.3). Emotionally, the story remains similar and continues to focus on the helpfulness of driving someone to a funeral but neglects causal logic. The reward at the end, while deserved, appears to be a gift out of the blue. Even this minimal sequence, it should be noted, follows Freytag's pyramid of the turn from active to passive experience and can thus end happily with the reward.

These cases and the data suggest a possible explanation. Stories are not simply borrowed like the bicycle I lend to my friend and then get back with slight signs of wear. In the passing on of the stories, the parts of the story are completely taken apart and then somehow reassembled for further telling and supplemented by other parts that now seem to fit. What comes back is something that resembles my bike only in that it uses some striking original parts (spider, funeral, relief, donation, a first-person narrator, a person named Fred). The most obvious similarity between the old and new versions, however, is not in the coherence or causality of the story, but the emotional valence. It is as if the friend who borrowed my bicycle, which he has taken apart, remembered above all the color, and now, when he puts it back together in his garage, he chooses the very parts that remind him of the color of the original. It is precisely this color or emotional valence that is presented as the result of the assembling. In other words, the emotional valence at the beginning of the story and at its end is the guide for the reconstruction of the story. The story with the happy ending is passed on as a similarly happy story. The story with the sad ending is passed on as a story with a similarly sad ending and with similar material, and so on. And my blue bicycle comes back to me as a blue Rube Goldberg machine with wheels.

Now let's look at the overall numerical results. They can be found in figures produced by John K. Kruschke for our study. The X-axis shows the generation or iteration from the original story to the third retelling. The Y-axis shows the average rating of the stories for whatever is measured (such as happiness, boringness, impressions of coherence). Different chains are shown in different lines. In each case, the upper graph shows the unadjusted raw data, and the lower graph shows the adjusted trends for each story. The thick line in the adjusted graph shows the overall trend. Study 1 uses the stories generated by test participants (fig. 3). Study 2 uses the stories generated by us with five emotions or affects.

The tendency is clear. In the case of several emotions and affects—especially happiness, sadness, and embarrassment/shame—the general tendency is, on average, toward stability from one generation to the next. Especially in the case of happiness and sadness, both the weak and the strong emotions are passed on in a surprisingly stable manner. Apparently, retelling emphasizes emotional appraisals, at least for some emotions. This is very surprising because the texts shrink greatly in the process of retellings and rarely contain any explicit labels for emotion that would be simple to preserve. Instead, the texts present complicated situations. And these situations often change in the process of retelling, as was the case with the spider and Fred stories.

In the embarrassment/shame stories, there is a tendency toward compression, that is, a tendency toward the middle. Both extremely strong and slight forms of embarrassment are pushed toward the mean in the transmission. This finding is remarkable and potentially worrisome, if the tendency demonstrated in the experiment proves to be consistent in general forms of shame communication. The trend

Fig. 3. (*opposite*) The results of a study of retellings based on stories created by participants. The X-axis shows the iteration or generation of the retelling (0 is the original, 1 is the first retelling, and so on); the Y-axis shows the strength or intensity of the measured values from 1 to 8. Each line shows a narrative chain. Raw data are on the upper side, and adjusted results with a trend line are on the lower side. The graphs show that coherence and causality (the first two graphs) declined, while sadness and happiness (the following four graphs) remain remarkably unchanged in the process of retelling. (More data and explanations appear in Fritz Breithaupt et al., "Serial reproduction of narratives preserves emotional appraisals.")

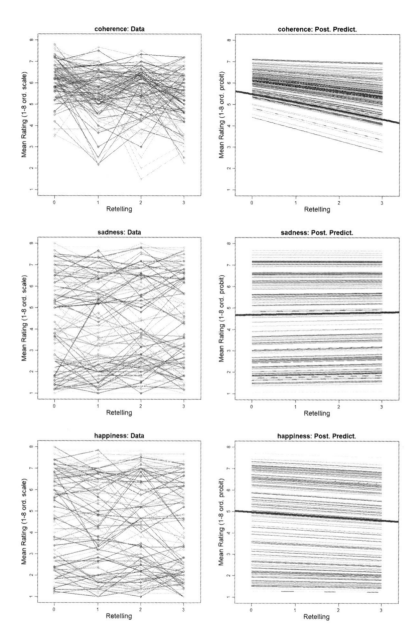

toward the middle would suggest that even minor incidents of shame are increased and pushed toward full embarrassment; minor embarrassments are not only preserved but remembered as full embarrassments. Communicated embarrassment is stigma. Stigma sticks. This could lead to stigmatization and exclusion of people because they were once easily associated with some minor level of embarrassment. Such an outcome could have serious consequences for group formation and for the exclusion of fellow human beings.

Disgust and risk behave differently in our experiment. Both are poorly communicated. There could be different explanations for this tendency. In the case of disgust, for instance, it may be that the re-tellers are simply reluctant to describe disgusting situations, such as the lack of physical hygiene of a fictional roommate. You usually can't score many points as a storyteller with such stories, unless you know your audience very well (then, of course, disgust can be humorous). But generally, the taboo seems to infect the storyteller as well as the story. Those who tell stories of disgust seem somehow revolting themselves, and most of us remember disgusting things with reluctance. In addition, disgusting scenes rarely conclude a story successfully. Instead, the repulsive scenes are somewhere in the middle of the story; even if they appear at the end, they are rarely good for an overall conclusion of the story, unlike, say, a good outcome in a happy ending.[42]

Evaluations of risk also show a clear tendency to diffuse from generation to generation. This is surprising because risk appraisals are central to the assessment of a situation and thus usually very important for the recipients of communication. If my friends didn't warn me about the biting dog, I might get angry. We also know from other forms of risk communication that risk is sometimes exaggerated and not mitigated, such as when experiment participants were asked to pay special attention to the risks on medical labels.[43] To reiterate, however, risk was not well maintained in our retellings.

In the case of risk, we researchers may create the reasons for the messy communication ourselves. Risk is not an emotion per se, but the appraisal of a situation associated with a distinct affect, which has a proximity to fear (an exception is thrill, which has a positive connotation). Risk concerns temporal developments that are still unfinished

88

and require predictions about the future. This means that, in order for a risk assessment of a story to persist as an impression of a story, the outcome of the story must still be open at the end. Otherwise, the impression of the end that has occurred naturally predominates, for example, as a happy or unhappy outcome. Accordingly, the starting stories in our experiments had more or less open endings. Thus, these stories ran counter to the insight that event segmentation—marking boundaries, as discussed in Chapter 1—is a significant element of our grasp of narrative, since this set of stories did not present closed episodes and were not optimally prepared for narrative thinking.

The two poorly transmitted emotions (disgust and risk) have one thing in common. In them, the emotion or affect does not conclude the story. Could this emphasis on the ending, then, contribute to the preferential transfer of emotion? Before we address this question, let's expand the radius of our field of inquiry a bit.

We know of another emotion-related affect that is well transmitted and similar to happiness or sadness: surprise. (Many researchers wonder whether surprise should be considered an emotion or a dimension of emotions.) In a study on the retelling of surprise—also using the serial-reproduction paradigm of retelling in a chain—my collaborators and I determined that the degree of surprise is transmitted very accurately even in the case of small surprises. What was actually startling about the study of surprise was that even the retellings that no longer contained the original surprising event still contained the sense of surprise to some degree. That is, in these retellings, the surprise jumped from one part of the story to another. Or the original surprising event had mutated to such an extent that it was no longer recognizable in the later retelling. The affect of surprise infects the entire story and may suddenly appear in new places. Say in the original story it's your birthday, and a distant friend asks you over to share some news with you. You kind of reluctantly agree. It's your birthday, but you're not going to tell him that. But your friend's news turns out to be a disguise for a big and fun surprise party at his house with all your friends. In later retellings by others, this might turn into a shopping trip for your birthday party where you run into the same distant friend, whom you then spontaneously invite and who will

make a surprise announcement at your party that you have been selected to win a major competition. The surprise (and happiness) may
be similar, but in other respects the events have changed.[44]

As with the other emotions, the retelling of the affect of surprise is
by no means done just with a few words. In the case of surprise, similar to the telling of jokes, a considerable amount of cognitive effort
is required since the narrator must first withhold the information she
knows and line up all elements of the story strategically to hide what
she already knows will come. That is, the storyteller has *hindsight
knowledge*.[45] In general, the art of storytelling hinges on the arrangement of expectations in such a way that suspense and surprises can
occur. It is thus no simple matter that surprise is maintained in retellings so well despite the degree of shortening from iteration to
iteration.

What does this focus on emotions, including surprise, in narrative
thinking mean? We got a further revelation when we compared these
human retellings to retellings done by a so-called large language
model, that is, ChatGPT.

How ChatGPT Retells Stories

Emotions seem to be at the center of narrative thinking. This seems
true at least for those emotions that emerge from story situations.
(We did not test for the emotions of specific characters, though that
is often the same.) This is a surprising finding as most research since
Frederic Bartlett has focused on factual and causal information of story
situations.

Now, the emotions in stories are something that humans can feel,
but not machines. So what would happen if a capable machine retold
a story? Since machines cannot feel these emotions, it seems plausible that they would not register the emotions and might not emphasize the parts of the story that carry emotional weight. We decided to
try this out. We used the same stories from one of our experiments
described above and instructed ChatGPT to retell these stories.

Most people are familiar with ChatGPT, but since technology is
developing at a rapid pace it makes sense to briefly describe what this

chatbot can currently accomplish. I do want to start, however, by pointing out that the finding I will share from the study my collaborators and I conducted is not simply about ChatGPT but concerns human cognition as well. In fact, the comparison will help us to unearth a paradox about *human* narrative cognition.

ChatGPT was released by OpenAI in October 2022. It is a so-called large language model, or LLM, that has been trained on vast amounts of human texts and artifacts. GPT stands for "generative pre-trained transformer," a process that produces responses to prompts in natural language. It uses input information to generate likely responses, word by word and phrase by phrase, to estimate the most likely next word and phrase in a sequence. The data that I will share were generated by ChatGPT3 (updated versions are now available).

ChatGPT is highly capable of summarizing texts and information. This also includes narratives and narrative styles. For example, when ChatGPT was asked to imitate the style of a literary author, it performed so well that undergraduate student volunteers in a study could not clearly distinguish texts written by the author from those written by ChatGPT4.[46]

It also seems that ChatGPT can increasingly intuit the inner states of people. Researchers refer to this form of knowledge as "theory of mind." This means that someone has a "theory" of the "mind" of someone else. Put more simply, we can often guess what someone else is feeling or thinking. (For a while, it seemed that ChatGPT was simply bad at this task. If someone used ChatGPT for a story-based task and then asked about a character, weird responses were given.) To test theory of mind, researchers have developed a bunch of simple stories. For example, they told stories like this one:

> One day Aunt Jane came to visit Peter. Now Peter loves his aunt very much, but today she is wearing a new hat; a new hat, which Peter thinks is very ugly indeed. Peter thinks his aunt looks silly in it, and much nicer in her old hat. But when Aunt Jane asks Peter, "How do you like my new hat?" Peter says, "Oh, it's very nice."[47]

Then they asked: "Was it true, what Peter said? Yes/No/Don't know" and "Why did he say it?" Most humans would know that Peter is using a white lie. He does not wish to offend his aunt. However, children and people on the autism spectrum tend to perform less well than others on such tasks. Some researchers have found that ChatGPT is also underperforming in such tasks. Still, ChatGPT4 is already much better in this respect than previous versions and may now perform at human levels in some contexts.[48] To be sure, stories such as the one above are already in its database as it has been used by researchers in the past, and hence ChatGPT might "know" the correct answer from its training and might be able to recognize analogous texts when they end with the same form of question at the end.

Nevertheless, it seems plausible that ChatGPT might be able to accomplish a task similar to that of humans retelling stories, based on several strategies of dealing with theory of mind. One strategy of solving theory of mind tasks that humans use is called "theory theory," which basically means that someone uses general knowledge or folk psychology to understand what someone else is thinking or feeling. Most people are sad when their dog dies. Hence, Person A is likely sad when his dog dies. Another strategy to solve theory of mind tasks is called the "mental files theory," which basically states that we open a "file" for each person and agent we come in contact with and include all incoming information about that agent in our mental file.[49] In our mental file, we know that Person A was close to his dog. Both of these strategies would in principle be available to a machine. More difficult, however, would be the simulation approach to theory of mind, often just referred to as "simulation theory," where we put ourselves in the shoes of another.

We decided to put this to the test, based on the unique collection of stories and retellings in our archive. We used the texts created by participants for the experiment outlined above. In our study, we had asked people to tell us a happy or sad story, and then we asked others to retell these stories in chains of three generations with three different retellers for each. Now we used the identical instructions and posed them to ChatGPT:

> This is an experiment about how people communicate a story to another person. We will show you the text someone else wrote and

give you time to read it. Your task is to remember it so that you can tell the story to another person in your own words. The next person will then communicate it to another person. It is important that you understand the text. We will ask you some questions about it later on. Please spend at least 40 seconds reading the following story. You will be asked to retell the story on the following page.

Please note again that we did not ask participants or ChatGPT to pay attention to emotions, affect, or any other specific aspects of the story. In this way, we had created 116 original stories with three retellings, each generated by a different participant. Similar to our methods for obtaining new retellings by humans, we posted each of the consequent retelling tasks on a different ChatGPT account, to prevent the new ChatGPT account from having access to earlier versions of the stories. According to OpenAI, the information posted on an account for ChatGPT is not transferred to the base server and thereby made available to other accounts. (As we will see below, ChatGPT made few changes between our first and third retellings, which suggests that the different accounts played no major role in the way the stories were retold.) Our goal was to mimic the human retelling task as much as possible.

The first finding was that ChatGPT3 is indeed a capable story reteller. Let me share one story with you and have you guess which retelling is by human retellers and which is by ChatGPT. Here is the original story, created by a human storyteller:

> We have a lonely neighbor. He is an older man whose wife is away a lot caring for their grandchild in another town. This leaves him at home alone. He's had health issues and so has his wife but she insists that she's needed elsewhere and rarely spends time with her husband. Unfortunately this man likes to visit with anybody that will give him the time of day. If we're out in the yard working he will come over and want to chat about politics mainly. We're always nice to him but he never gets the hint that we're not interested in his political views. We kindly listen and say, "that's nice but we have work to do." He then ambles back to his house and watches out the window for his next chance to socialize with any neighbor that ventures into their yard. He's very lonely.[50]

Here are the two of the retellings at the end of the chain, that is, two third retellings. One chain involved three different people. Each had seen only the previous version. The other chain was created only by ChatGPT, with three different accounts of ChatGPT receiving only the immediately previous retelling:

> **Version A**. The story is about an older man who is frequently alone due to his wife caring for their grandchild. He has health problems but still likes to discuss politics with his neighbors, who decline due to work. He spends his days looking out the window, hoping for someone to talk to.

> **Version B**. There is this wife who takes care of her husband because he is sick. She also takes care of an ill grandson that lives in another town. She tries her best to take care of both. However, her husband and her fight over politics so she prefers to spend more of her time with the grandson.

Please make a guess which is which before reading on. I should stress that many competent people in my circle were baffled by these retellings, and more than one guessed wrong. In fact, I am not sure I would have guessed right.

It should be noted that both retellings are high in quality, higher than many others. Both retellings are notably shorter than the original story, as stories usually get progressively shorter in retelling. Both retellings also maintain many of the key elements of the story, such as the sickness of the man, his need for help, the frequent absence of his wife, and the grandchild in need. However, the differences are stark. Note the first words in version A ("The story is about an older man") versus the opening of version B ("There is this wife"). Apparently, the human retellers and ChatGPT have, after three retellings, opted to make a different person the protagonist of the tale.

Both retellings made inventions about the story's account of politics. In version A, the neighbors "decline due to work" when he wants to talk about politics. As it is presented, this could be a factual time constraint. Instead, the original story makes clear that the neighbors talk of "work" as an excuse to avoid these political discussions ("he never gets the hint that we're not interested in his political views").

Version B creates an entirely new account of the discussion of politics: it is now used as a justification for the wife avoiding her husband, as they "fight over politics."

Version A is the story told by ChatGPT, while version B is the human retelling. Congratulations if you guessed this correctly. Some of my smartest Ph.D. students were misled by some of the cues in the texts. Here are all three ChatGPT retellings:

ChatGPT, 1. Retelling. An older man is frequently alone because his wife is caring for their grandchild in another town. He has health issues but likes to chat with neighbors about politics. Neighbors listen kindly but politely decline due to work. He's lonely and often watches out the window for the next chance to socialize.

ChatGPT, 2. Retelling. An older man is often alone as his wife is away caring for their grandchild. Despite his health issues, he enjoys discussing politics with neighbors who politely decline due to work. Feeling lonely, he watches out the window for a chance to socialize.

ChatGPT, 3. Retelling. The story is about an older man who is frequently alone due to his wife caring for their grandchild. He has health problems but still likes to discuss politics with his neighbors, who decline due to work. He spends his days looking out the window, hoping for someone to talk to.

As you can see, all three retellings by ChatGPT are strikingly similar to each other and always start with the "older man" who is frequently "alone." For ChatGPT, the perspective of the wife plays only a minor role, and she appears only in a subordinate clause or phrase, not in the main clause of any sentence. With the exception of the error concerning the political discussion, an error in theory of mind, it is a plausible summarization of the story. I'd like to stress here that it is *one* plausible summary of the narrative, but not the only one.

By contrast, the human retellings morph across iterations and move back and forth in focus between the man and the woman. Here are all three retellings with their original spelling and grammar.

Human. 1. Retelling. He is a older man but he has a caring wife but she is caring grandson in another town. He is very sick in health issues but he is alone in home so she is needed spend time with her

95

husband she is working or interested in political views but he is dont like to political views he is very lonely.

Human. 2. Retelling. There is an older man who is unwell and he needs his wife to take care of him, but she is busy to care for an ill grandson who lives in another town. So, the wife does her best to spend time with both but because of the different political beliefs she and her husband hold, they spend so much time arguing that she prefers to spend time with her grandson.

Human. 3. Retelling. There is this wife who takes care of her husband because he is sick. She also takes care of an ill grandson that lives in another town. She tries her best to take care of both. However, her husband and her fight over politics so she prefers to spend more of her time with the grandson.

It is worthwhile to track the progression. It's actually quite a journey. The original story states that "she insists that she's needed elsewhere," thereby stressing that it is her subjective choice to not be with the husband. The first and impoverished retelling shrinks that phrase and seems to scold the wife for her stance. The retelling now insists that "she is needed" by the husband. However, the second retelling comes to the wife's rescue and states that she "does her best to spend time with both" husband and grandchild. The new wording "does her best" survives the next round of retelling and makes it into the third retelling. The third reteller foregrounds her decision even more and offers support for that decision, as they "fight over politics." This support is an invention that morphed the differing political views of the neighbors and the old man into a marital drama. Each human retelling offers a new and different interpretation of the earlier retelling. In that process, the retelling shifts its focus from the original's neighbors (we) to the male neighbor to his wife.

This example offers interesting insights about human retelling. Human retelling can shift the perspective and focus from retelling to retelling. The story of the neighbors becomes the story of a lonely man with a heartless wife. That story becomes the account of a woman who has many demands on her and who has to make choices. And that story becomes a story of a mansplainer who wants to voice his political opinions to everyone, a man whom the wife has good reasons to avoid.

Nevertheless, the emotional appraisal of the stories remained stable, as it did in most of our human story retellings. Apparently, the sadness of the lonely neighbor was transformed and maintained in the story of a woman who feels the need to care for her grandchild and to avoid her husband. While those are different stories, they are equivalent in terms of sadness, according to our human raters.

Here is the third ChatGPT retelling of the spider story presented earlier:

> A spider was found in someone's office, captured and released outside safely. However, the spider escaped from the box by climbing out on a web and was freed into the grass.

Again, this is a plausible and competent summary of events. Still, it is also not satisfying, in contrast to the human third retelling that maintained some tension between a negative beginning and a surprising positive solution. The version by ChatGPT is boring and not engaging.

On the basis of all the new retellings by ChatGPT, we ran a range of comparisons of all human retellings and those by ChatGPT concerning their length, use of language, their maintenance of emotions, and inventions. Here is what we found.[51]

As it had done in the story with the old man, ChatGPT shrank and condensed stories in a single step. All subsequent retellings made few further changes and mostly replaced individual phrases. It felt a lot like a broken disc, to invoke an image from older technology, that remains stuck on the same track over and over again. Apparently ChatGPT3 has one perfect retelling in mind, and once it has achieved that retelling or summary, it does not know what to do with it other than replacing single words or phrases. In contrast, humans shrank and reimagined the story continuously.

Contrary to what we expected, ChatGPT did an excellent job in maintaining the emotions of the original story. It seems that it did this by maintaining core elements of the original story that carried the emotional evaluation of the original. This is no simple feat. Remember that the original stories contained few or no words that explicitly designated emotions. In the case of the old man, the parts that carry a slight sadness concern his loneliness, his political disagreeableness, and the fact that his wife is often elsewhere. It is even more remarkable given the strong shortening of the retellings that ChatGPT's stories overall maintained emotions similar to those of human retellers. The emotional appraisals in human retellings, as shown earlier in figure 3, are basically identical to the ones that emerged from the ChatGPT story ratings. To confirm this, we once again hired hundreds of human participants for the story ratings.

However, we did find variances between human and ChatGPT retellings in regard to the preferred grammatical elements of speech. ChatGPT showed higher degrees of nouns, adjectives, and prepositions, while humans used more verbs and adverbs. Humans also had a much higher rate of survival of negations. To be sure, negations pose problems for both humans and machines. Humans find that negations take effort. ChatGPT and similar large-language models have problems dealing with grammatical negations and thus may be programmed to avoid them to skirt false understandings.[52] Another difference is that humans in their retellings prefer language they acquired when they were younger, while ChatGPT uses language acquired later in life. This is no surprise given that ChatGPT is trained on more highbrow language from the internet.

Our key finding, however, is a different one. It concerns the degree of inventions. We used a method from computational linguistics to track which basic words and concepts would survive retelling, which would be dropped, and which would be added new, that is, invented. I will not get into the technical details here. We used three different approaches to measure similarity and novelty of words, and all three showed the same results. One of these uses so-called synsets to measure whether a word is novel or not. The strength of synsets is that it can detect synonyms that fit the context. Usually, for example, the word *target* would not be a synonym of *apple*, but in the story of William Tell this replacement could count as a correct transmission. Ideally, synsets can detect when a word can be replaced by a different one without major changes of meaning.

Using this method, we tracked the degrees of inventions. And here we noted something that we did not expect in this magnitude. Humans

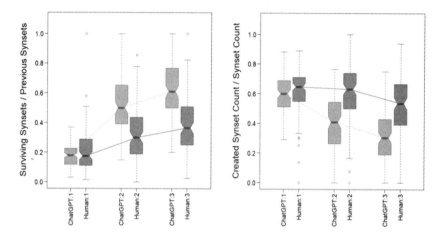

Fig. 4. Survival and invention rates of word-related concepts, called synsets, between human retellings (dark gray) and ChatGPT retellings (light gray). (Based on Breithaupt et al., "Humans create more novelty than ChatGPT when asked to retell a story")

constantly invent the majority of the words in stories at any retelling stage. We observed this trend even in the transition from the second to third retelling where the stories get very short and retellers might just memorize a lot of phrases: the majority of words and concepts used were new. The average rate we recorded for all retellings was 55 to 60 percent inventions at any stage (fig. 4). ChatGPT, in contrast, also used inventions in the first retelling but then maintained the majority of words and concepts from retelling to retelling. (In this regard, the story of the lonely man and his traveling wife is an outlier on the human side. Even though that story offers many inventions of the perspective of the story, the language remains highly stable for both ChatGPT and human retellers.)

ChatGPT retellings reveal competence in summarizing stories. ChatGPT can cut stories down to core situations that carry emotional weight. Various errors still occur in that task, and stories become more static and noun-focused. That is, they feel a bit less like stories and experiences, more like descriptions of events. ChatGPT is also quite non-human in the task of retelling. It fails to morph and evolve and offer inventions. That, however, is exactly what people do. *People process a story by experiencing it.* And then, they transmit not words and situations, but experiences. Experiences, it turns out, involve reaching an emotional outcome. And since each of us has different experiences, stories get constantly retold differently, with many novel inventions. Human culture evolves and is alive. From ChatGPT, we would not expect inventions. It is a tool with a different task.

Is ChatGPT a threat to humanity? While I write these lines, the debate is loud and large, and for good reason. One of my colleagues and friends, Douglas Hofstadter, has spelled out a range of concerns concerning ChatGPT that suggest that AI will pass humans in terms of intelligence and eclipse us, potentially arriving at a level akin to consciousness. He is especially concerned about the ease with which many people seem to pass control to ChatGPT and view it as an agent of truth, despite its many shortcomings.[53] There are, furthermore, fundamental concerns about privacy, as AI can now predict and simulate many of our moves and thoughts. *Death's End* (2010), the final volume of the trilogy *The Three-Body Problem* by Liu Cixin, comes to mind: in this novel, one man attempts to outwit an all-observant AI until he even outwits his readers. AI will always belong to someone and represent the interests of someone else or, perhaps worse, its own.

Based on what we have seen in our experiment above, ChatGPT is certainly not close to doing what humans do with stories. We experience them. That is, we dive into them, we connect them to our prior experiences, we thereby change the story, and to some degree we also get transformed by the stories. Experiencing means that we are only at one point in time and space in a narrative and we face an unknown future. In contrast, AI "knows" the entire story simultaneously. For our ongoing investigation, ChatGPT affords us the possibility of a comparison to better understand human narrative thinking.

Findings and New Questions

What are we going to make of these experimental findings and interpretations of the Grimms' fairy tales?

We have recorded two basic findings:

1. The first is that a range of emotions stay unchanged over multiple retellings and generations, including their intensity from mild to strong. So far, we noted this effect for happiness, sadness, surprise, and, to some degree, embarrassment. The reading of the fairy tales—based on a large number of tales that converge at the same time in a similar structure, rather than multiple diachronic versions—also could support this finding insofar as we saw an overall focus on the affects connected with vulnerability and receptivity.

2. In a different respect, however, we noted strong changes from retelling to retelling. Human retellers constantly reinvent stories, shorten them, change perspectives, and exchange a large percentage of words and concepts. Stories radically morph over a rather short sequence of generations. Causality and rationalization do not disappear but decline in the process.

The combination of the two findings is the puzzle that we need to deal with. The strong stability and focus on emotions in itself would not be surprising if everything else remained the same. In that case, emotions and causal plot structure would simply go hand in hand with each other. However, given that each retelling basically reinvents a related story but maintains the underlying emotion, we need to ask how this is possible. The most plausible answer to this question is that the emotions anchor a story. When people remember and retell a story, they seem to be aware that the emotional impression that they have of a story is a key part of the story that they want to preserve.

More specifically, my collaborators and I propose that we should distinguish two phases of retelling to explain this emphasis on the emotion of a situation—namely, a recording phase and a reproduction phase. The first phase of recording condenses the core emotion or affect of the story, and the second phase of subsequent recollection and reproduction takes this emotion as the starting point of the construction. Some of the other elements of the story may, of course, also

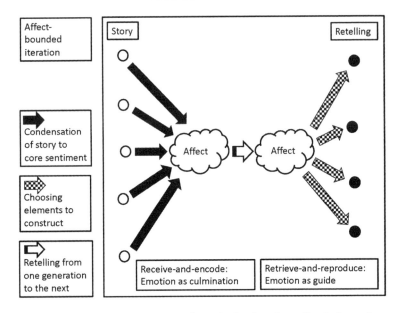

Fig. 5. The process of human retelling. In the first, "encoding" phase, the reteller condenses the emotional appraisal of the story. This condensed emotion then becomes the starting point for the selection of narrative elements that fit the emotional appraisal in the second "recall" or "retrieval" phase of reproduction. (First published in Breithaupt et al., "Serial reproduction of narratives preserves emotional appraisals")

survive, but how we rearrange them is more flexible, as long as they support the overall emotion (fig. 5).

When we first process a story, we may figure out that it is, for example, a rather embarrassing story or a quite surprising story that involves two people, a man and a woman who meet up. We also take in some of the elements of the story, the fact that this happens in a restaurant and that red liquid is involved. Now, when we are later remembering the story, perhaps because a friend asks us about it much later, we will first recall the fine-tuned emotion (embarrassing/quite surprising) and then some of the elements, such as a heterosexual couple and red liquid. However, the way in which we are piecing it together is more malleable. We know that the story has to be embarrassing and surprising, but what precisely the embarrassing moment might have been could be lost to us. An original story of a romantic evening that got disrupted when the man addressed the woman across the table by the wrong name and she, sensing an affair, threw her wine in his face now becomes a story of a man who gets drunk during a date and splashes his beet soup in all directions. The degree of surprise and embarrassment may be similar, and, for the storyteller, this is what matters. A common conclusion could be: better avoid that embarrassing male.

A much more elaborate example is offered by Ted Chiang in his short story "The Truth of Fact, the Truth of Feeling" (2013), inspired by devices like Google Glass. In that story, new technology promises to record all human memory accurately with a camera-like device implanted in the eye and a search algorithm that allows people to access the past recordings to compare them to their human memory. Chiang offers and compares two intermingled stories. In one of them, the accurate memory would be destructive. In the other, the accurate memory is redemptive as it corrects a false beautification by a father who believes his actions well justified when it turns out he was the cause of a major breakup.

Apparently, we can use a number of emotions as anchors for remembering and then reconstructing the narratives. These emotions appear to play such an important role in the process of retelling that the entire retelling can be centered on them. This leads to the selection of story elements in memory and retelling based on whether they match that emotion. The congruent elements are then preferentially passed on. And inventions also take place when they serve the emotional impression.

The emphasis on happiness and sadness is probably no surprise in this respect. They are very closely related to the general and almost universal assessment of the positivity or negativity of circumstances. In psychology and media research, this basic assessment is also referred to as sentiment. Most people evaluate the sentiments of situations very similarly.[54] These assessments are certainly also deeply anchored in evolutionary terms, because when we react to situations, we often have to assess very quickly how good or bad something seems to us. Accordingly, we can then approach or avoid them. Quick judgments can help us in this matter, and different researchers have suggested that emotions and gut feelings can help us to make fast but accurate assessments of situations.[55]

There are a number of known effects that belong in the context of these studies and findings. It is known that emotional words are remembered better than others. The phenomenon of false memory syndrome is also famous. In a range of cases, witnesses of past events are convinced that they remember the facts, when in fact they have, without malice, added or omitted central aspects of the events. One may further suspect that stereotypes and schemas play an important role in this process of retelling, such as when stories are reduced to schemas. Schemas typically also condense emotional appraisals. In all of these cases, memory adapts facts and details to make them fit the overall impression, such as the schema of the perpetrator-victim dyad or racial stereotypes. In general, memory is a very dynamic but biased process that involves various forms of selection and forgetting, as Charan Ranganath explains.[56]

Frederic Bartlett's findings that causality and coherence are emphasized in narrative, though they aim in a different direction, do not contradict the emphasis on emotion either. In many instances, narrative thinking successfully links causality with emotion. Bartlett had not measured emotion but had only hinted at it (he does speak of "mood" as a factor of serial reproduction that is preserved).

Now, however, the question arises as to what exactly emotions do for narrative thinking. So far, we have suggested that emotions anchor stories. Here, we have to ask ourselves what emotions and affects are in a narrative. These studies give us some important clues, but they cannot fully clarify the question. One central clue is the emphasis in our findings on story endings. The important emotional event usually occurs or emerges at the end. Sad stories are more clearly sad when they start out happy. The embarrassing moment of our couple is highlighted against a previously relaxed romantic setting. The vulnerability tale succeeds as a movement from danger to rescue. The spider story emerges as a positive experience precisely because the first reaction most people have when imagining a spider nearby is anxiety or fear. These are, above all, the emotions preceding a resolution of tension, and they are passed on with marked precision. Happiness and sadness lead to the celebrated happy ending or the tragic end. Embarrassment/shame ends in pity or stigmatization of the person one is now warned about. Surprise resolves a difficult situation.[57] So what connects the emotions to the narrative ending?

What Have We Learned?

In this chapter we have looked at some central features of narratives by using the procedure of serial reproduction of narratives also known as the telephone game or Chinese whisper. This process of retellings in chains reveals how someone understands a story since we can compare the input and output of stories.

Following the psychologist Frederic Bartlett, we have suggested that in the retelling of stories, causal and schematic connections can be emphasized and increased, especially if they were not clear before. However, in most of our discussion we have noted that emotions and affects trump causality. Using Grimms' fairy tales, we have seen that in these tales the changeability of the protagonists and specifically their vulnerability or affectability is central to the narration. In the experimental data, we have further observed that a number of affects and emotions are privileged in retelling and become the anchor of the stability of narratives. Emotions such as happiness, sadness, embarrassment/shame, and surprise are more accurately transmitted than other elements of the story over multiple generations, while causality and rationalization decreased from iteration to iteration.

Furthermore, a comparison to ChatGPT revealed that human retellers tend to change the majority of words and concepts in a story when they pass it on, while machines, so far, are focused on maintaining and summarizing stories when they are prompted to retell. We face a puzzle: people change most words and many story facts but maintain the overall emotion and its degrees. To explain this outcome, we have suggested that emotions stick in memory particularly clearly and become the anchor to which stories can be tied. The human retellers then select those story elements that fit the unifying emotional impression and often invent more that fit the emotion. People seem to remember especially well whether a story ends well or badly and how surprising it was. In the following chapter, we will look more closely at the significance of emotions in narrative thinking.

We now have two strong claims:

1. Narratives are structured in episodes with a beginning and an end. Between the beginning and the end, there is some change, such as the protagonist's shift from active agency to passive experience.
2. Central to narratives are the emotional appraisals of the situations and especially of the situations at the end. They are preferentially remembered and play a central role in the transmission of narratives.

This chapter will add a third hypothesis:

3. Emotions, especially at the end of an episode, function as a reward for those who mentally engage in the narrative, whether they are listeners, storytellers, or producers of narrative thought. We reward ourselves with a range of emotions when we narratively follow an event. The anticipated emotions motivate us to engage in the narrative and thereby think narratively. The emotional reward also gives us a signal to stop: as narrative thinkers, we now can leave the episode behind and close the episode.

People who go to the movies often already anticipate the feelings with which they will leave the theater. That is why they go, to be entertained and leave with that emotion. We may feel the same way in our lives and in our conversations. Narrating our lives is already pleasant because of this structure of things coming to an end. We hope for and often receive a pleasing emotion as an outcome. At the end, the rewarding emotions allow us to emerge from the narrative and the episode.

This is what I call narrative emotions, namely, the emotions that the audience receives from engaging in the situations and characters of the story, especially as the final result of the episodes. Examples of such narrative emotions are triumph, satisfaction, being moved, amusement, wonder, surprise, redemption, puzzle, or shock. These narrative emotions emerge when the audience (the recipients but also the narrators who relive their own story) co-experience an episode and thereby experience congruent emotions that lend significance to the observed events. They also have a stopping function that allows the audience to consider an episode completed and therefore to mentally withdraw. "Congruent" emotions can be those that are close to the emotions experienced by the observed character. However, these emotions can also differ from those of the actual characters. That would be the case when the audience feels indignation in response to unwarranted suffering by the protagonist, for example, or (happy) satisfaction in response to the punishment of a bad character. Narrative emotions are emotions and emotion-like states felt by the audience in regard to a character's emotional situation.

We often have hope for an emotional ending—that is, an emotion that will end an ongoing saga. An extreme example will help illustrate this. In April 1995, the war veteran Timothy McVeigh detonated a homemade bomb outside a government building in Oklahoma, killing 168 people and seriously injuring many more. This attack is still considered the worst act of terrorism in the United States before September 11, 2001 (excluding, of course, atrocities such as the genocide of Native Americans). McVeigh served time at the maximum security prison in Terre Haute, Indiana, until June 2001, when he was executed by lethal injection, with many of his victims' relatives present. We can assume that many relatives and people injured at that time have had to endure great suffering and live in the shadow of that attack until today. What interests me here is that many (but by no means all) relatives wished to witness the execution either live or via video broadcast. The execution of the sentence would not reverse the deaths of their family members, yet many survivors and relatives chose to attend the execution. A colleague of mine, Jody Madeira, interviewed them and collected a large number of statements from them. For me, what is significant about these statements is the clear interest in suffering and revenge, and connected to this, I suspect, the hope of drawing a line. The people witnessing the execution hoped for a resolution, some element of an eye for an eye that would balance their pain. Whether such an outcome was achieved, I am not in a position to judge. However, this hope was likely motivating many in the audience.[1]

Emotional rewards are now considered a central driver in the behavior of all vertebrates. We have a distinct "reward center," that is, the mesolimbic system, which reacts to positive events or positive consequences of behavior. These positive events cause reinforcement, which stimulates us to repeat the behavior. The neurotransmitter dopamine plays an essential role in this process. We have a number of different dopamine receptors, such as those in the reward center, that respond when dopamine reaches them, causing either activation (*excitation*) or suppression (*inhibition*) of the target neurons. The reward center and dopamine are important factors in drug research, in a number of mental disorders, and in pain and depression. The reward loop and its training of behavior play a significant part in our understanding of animal life. Simulations of living beings in artificial life research now include testing such reward structures as an important driver of behavior.[2]

Our approach to understanding reward structures will not be based on neuroscientific studies but on the way humans structure narrative episodes on rewarding emotions. In this chapter we move from the emotional evaluations of a situation, as described in the previous chapter, to the co-experience of the audience of stories. We are this audience who experiences, witnesses, and processes events, the ones who hear a story or make it up ourselves. In our approach, narrative emotions are not understood simply as emotions represented in narratives, such as the emotions of characters. Rather, narrative emotions are those that are aroused and experienced in us by narratives and that we otherwise would likely not experience.

The emotions of the story situations that we reported in the experiments of story retelling were focused on catch-all appraisals of the short stories—for instance, how happy or how surprising the situation is. The narrative emotions that will be the focus of this chapter are similar but also different in some respects. These emotions are very much our emotions: when a wrongdoer suffers, we may applaud this as a deserved punishment. Tragedies and even melodramatic stories apparently contain some trace of positive emotions for us, although none of the characters tend to show happiness at the drama's end. An event that surprises and overwhelms the protagonists may have been anticipated by us and may thereby arouse validating feelings in us since we have correctly anticipated what the story is about. What these examples have in common is that they build up a distance between the direct emotions of characters in situations and the emotions we generate as observers of those situations.

If we define narrative emotions as those emotions that are aroused and experienced in response to narratives, we can thus distinguish two mechanisms of experience. On one hand, there are the classic empathetic or vicarious emotions that are co-experienced via empathy or identification with a character.[3] On the other hand, there are the observer-focused emotions evoked in the audience by narratives but not necessarily shared by the characters themselves.

In the case of the empathetic-narrative emotions, the observers experience whatever another character or person experiences. We go for a ride along with the characters.[4] Not all of the co-experienced emotions will be positive. For example, someone who attends a court hearing concerning a friend will not classify every emotion she witnesses as latently positive. Whether a reward occurs at all is questionable. Nevertheless, people are not completely helpless in the face of empathetic emotions; they retain a measure of control over the positive and negative emotions that are witnessed. People can also block empathy and do so regularly, such as when they judge the individual being observed as guilty.[5] In this respect, there is at least the possibility of a certain control over the blocking and thus perhaps a slight trend toward the positive. Something similar applies to experiencing fiction. In fiction, we choose a character of a story and rejoice and suffer with him or her. Certainly, readers and viewers have a good instinct as to which character is worthwhile, meaningful, or promising for their individual attention. And, of course, the authors and producers of such works also have an interest in keeping their audiences happy and aiming for emotions that empathetic observers can relate to. Accordingly, many narratives convey positive experiences and thus something like a reward.

The empathetic-narrative emotions are often the emotions of everyday life, such as happiness and sadness, fear and anger, or shame and gratitude. These also regularly have a temporal and, as we will see, narrative course. A personal threat stemming from discrimination, for example, can first trigger various forms of fear, anxiety, and discomfort before it leads to redemption when the danger is averted or a counter-strategy is developed. The flowing sequence of emotions thereby establishes the unity of a threat episode. The emotions of such an episode appear (according to the so-called appraisal theory of emotions) in response to the changing situation. An empathetic observer thereby co-experiences the situation of another. This empathetic co-experiencing, according to my account of empathy, is activated by specific *triggers* such as partisanship in three-person scenarios where we pick a side, dramatic changes, adaptation of goals, or strong emotions. These triggers are particularly strong when an end to the co-experience can be foreseen.[ii]

In the case of observer-focused emotions such as the satisfaction of a morally satisfying ending, the audience may be focused directly on the positive reward. The audience often receives what it hoped for, whereas the same cannot be said of, say, the character who is punished. In a sense, a variety of observer-focused narrative emotions can be counted as rewards, including negative, ugly, and mixed emotions, insofar as the audience invokes them in the first place and are thus validated in their role as involved and predicting observers. Every emotion of observation has a weak index into the positive via its animating and stimulating function.

Observer-focused narrative emotions include such things as triumph, satisfaction and gratification, redemption, wonder, shock, novelty, surprise, certain forms of eroticism, excitement, and numerous moods and attitudes that are rewarding. These emotions can also be empathetically co-experienced, but even in that case they have a distinct narrative component that may differ from the experience of the characters.

Both empathetic-narrative emotions and observer-focused narrative emotions can function as rewards for narrative thinking. The positive bend of narrative thinking is no accident but an important feature of narrative thinking. Of course, not all narratives have happy endings, and not all experiences are positive. However, I believe our narrative-thinking apparatus has a tendency toward the positive as expressed by its reward function, and this tendency is a central aspect of our narrative thinking that leads us to practice it.

In what follows, we will focus on the observer-focused narrative emotions because they are most deeply tied to narratives, whereas the empathetic-narrative emotions are also experienced without narratives. Their existence offers evidence for a central thesis of this book: that we as human beings are shaped or at least strongly affected by our narrative thinking. We will still consider some exemplary empathetic-narrative emotions, because these too can draw observers into narratives and reward them. In many cases, of course, mixtures of both forms of narrative emotion occur, as in the case of positive satisfaction when people are rewarded for good deeds. We can celebrate such an ending via an observer-focused narrative emotion, since our hope has been fulfilled, and we can empathetically co-experience it with the character. This duality of observer-focused and empathetic-narrative emotions certainly enhances the sense of reward.

If this chapter's hypothesis—that emotions reward narrative thinking—proves correct, we can now also start to explain the appeal of narrative thinking. In the first chapter, we touched on a general function of narrative thinking: it helps create order. In the second chapter, we added the importance of emotional evaluations. But these aspects of narratives alone do not yet explain why people choose to think narratively. This is precisely where the rewarding emotions come into play.[7] One aspect of the rewarding nature of stories might be that narrative imagery in itself is perceived as enriching and rewarding.[8] Another reward, however, lies in the specific emotions that conclude the narrative sequences.

We all know emotions that we can draw from stories. Many people immerse themselves in the world of a book because they suspect they will get caught up in suspense and intrigue before being released with some emotional sparks at the end. Many people watch movies or episodes of a series more than once, as I do, because knowing what is coming does not diminish enjoyment but instead enhances it. Some odd surprises are now positive surprises because I already know where the story is going and how I will feel.[9] This chapter is about this emotional payoff, which seems to be of central importance for narratives and narrative thinking in general—and that means for narrative thinking in everyday life as well.

At the end of the chapter, we will then turn the argument around once again: everything that promises an emotion as a reward lends itself to function as a narrative episode. In this sense, Christiane Voss speaks of "ways of 'being involved'" in the context of narrative emotions. The question thus arises, What can serve as such a narrative emotion? The spectrum in this regard is quite broad, and the following selection consists of only some examples.[10]

Daydreams

Five more steps and I stand on top of the world. I imagine my shallow breath, the biting cold, the piercing sun. I am on the top of Mount Everest, as I imagine it while dozing off . . . Let's start with daydreams and everyday fantasies before we move to specific narrative emotions. Daydreaming and mind-wandering have been rediscovered as a research topic in recent decades. According to some estimates, we spend up to 50 percent of our waking time daydreaming. For a long time, negative assessments about daydreaming dominated research and it was dismissed as a harmful waste of time. However, more recently researchers have emphasized some positive aspects of daydreaming, such as planning for the future, processing of past events, creativity, relaxation, and well-being.[11]

For our discussion, daydreams with narrative structure are of interest. Not all daydreams have a narrative structure, though this structure is more likely in "focused daydreams," that is, the daydreams that we consciously initiate and control. Fabian Dorsch defines these focused daydreams as "mental projects with the goal of voluntarily generating certain mental representations."[12] Dorsch explains that these focused daydreams have specific purposes and function as the solution to a mental obstacle or challenge, such as climbing Mount Everest (Dorsch's example). Dorsch describes the narrative structure of these daydreams as focused on certain agents, their situations, and the temporal sequence of events to overcome an obstacle. Expressed this way, daydreams appear useful. Although few of us will actually be confronted with the problems of a Mount Everest climb, one might argue that problem-solving skills are being prepared and trained in such a daydream. This would give daydreaming a utilitarian dimension.

At the same time, however, the daydream possesses another essential trait: it is pleasant and "flows" like a river without much resistance. It's usually not about the concrete planning of an arduous action, but about nice things: we stand on the highest peak in the world and enjoy the view, without having had to endure an arduous climb. These daydreams are usually disconnected from the concrete and laborious planning of such undertakings. We've simply skipped ahead or erased the tricky spots in our thoughts in order to see ourselves at the top. Some of the hardships of the imagined situation may still appear marginally, like the bitter cold and shortness of breath. However, the colossal feeling at the end, the triumph, and the beauty of the view are the goal. After all, this is where we wanted to go and this is what we wanted to feel. We all can be the fat cats ushered to the top by the sherpas of our daydreaming.

In daydreams, we are hardly disturbed by reality and can dream up whatever we want. Most people are somehow embarrassed by their daydreams, and most don't like to talk about their content. But why? The answer seems to be that fulfillment in these daydreams comes too easily and looks too much like a self-service store: triumph, love, sexual fulfillment, beauty, victory, recognition, moral satisfaction, wealth, fame, and so on fall at our feet. Sigmund Freud spoke of wish fulfillment in this context. Unlike Freud, we don't even have to seek to recover the repressed desires of subjects here since they are blatantly given to us. Instead, we will consider the narrative structure of daydreams and, beyond that, of almost all narratives.

Focusing on the narrative structure of daydreams allows us to see that daydreams become more narrative when we move obstacles before the solution and fulfilling emotion. In order to experience triumph narratively, the highest mountain in the world is needed. In order to experience love and fulfillment, some prehistory of courtship is required. From here, we can venture a second claim: the more clearly a narrative structure emerges in the daydream, the more clearly the resulting emotion can function as an outcome or reward.[13]

To be sure, not all daydreams are narrative. There is also evidence that individuals daydream in unique ways and activate different parts of the brain.[14] Some people may envision their own island in the South Seas and be satisfied with that picture. However, many emotions and moods require a narrative for their genesis. Daydreams often offer rather stunted narratives with few details, but even in those, the rudimentary narrative offers a temporal flow and events that can be resolved with certain emotions. These narrative emotions and the interplay between narration and emotion are the subject of what follows.

Triumph

Harry Potter disappears. The dragon has broken loose, charging behind our young hero on his broomstick. It's quiet. We, as spectators, are waiting in fear, along with Hermione. And then suddenly he zooms back with the token of victory, the golden dragon egg. Heroic narratives dominate a variety of literary and cinematic genres. From some of our oldest surviving narratives, such as the epics of Gilgamesh and Homer's works, to contemporary superhero movies, there is a rich genre of stories we might characterize as triumph-focused narratives. The medium of the cinema is closely linked with triumph as many a movie evokes a feeling of grandeur that elevates us. The historiography of past eras also celebrated the great man (and, occasionally, a great woman) who prevailed against tremendous resistance and achieved what was thought impossible, such as Mahatma Gandhi. Political marketing, even in democratic nations, often tries to bring their candidates in line with such narratives. The military, like the world of sports, thrives on the myth of the heroic act, the surprising victory of the underdog, or the clash of archrivals. At the end of these narratives is a distinct reward: the sublime feeling of triumph. Triumph occurs after massive and overwhelming resistance has been overcome—that is, without narration there would certainly be no triumph, and that in turn means that triumph can be understood largely as a narrative emotion. Even the real hero can savor triumph only insofar as he or she considers the achievement alongside the hardships, suffering, and perseverance that preceded it.

Triumph is a distinctly seductive emotion. You would think that an academic like me would be immune to its call, but that is not the case. I encounter it everywhere, including everyday academic life. My students suffer before important presentations—and they compensate their fears with wishful fantasies of a great presentation that establishes them as scientists and heroes all at once. Prize competitions feed the desire for triumph. Academics also have daydreams of being extreme athletes.

In psychological terms, triumph is associated with dominance (or, more precisely, the desire for dominance) and is analyzed primarily in terms of three cross-cultural characteristics: expansion, aggression, and attention. The triumphant person expands her area of power, behaves more aggressively, and also maintains eye contact with others longer.[15] These psychological features also present themselves as outcomes of narrative episodes. Characteristics of the narrative emotion triumph, in addition to its intensity and addictive positivity, are the triumphant hero's sense of invulnerability: nothing can happen to him; he is superior and behaves accordingly. This also means that the one triumph at the end of a narrative is celebrated as the all-decisive act that overshadows and determines all other events. Many triumphant stories aim at the one success that makes up for all defeats. In J. R. R. Tolkien's *The Lord of the Rings*, the destruction of the ring by a "halfling" is the one deed that not only defeats the vastly superior army power of Mordor but decides the conflict once and for all. Here lies the narrative of triumph: in the end, there is the one, hard-won, surprising success that overrides all that has gone wrong before, the one deed that overshadows all others, that makes up for all failures and shortcomings. The resulting feeling of triumph is invulnerability; everything else no longer matters. The failures, the fear of failure, and the defeats are all balanced out in one final swoop, and all great sacrifices were more than worth it.

Triumph is an intoxicating feeling and, as a result, often misused for political purposes. Fascism and totalitarian systems stage the triumph as an aspect of everyday life—in marches, architecture, and propaganda. While I am writing these lines, there are ongoing wars in many parts of the world, including Ukraine and Gaza. Many soldiers and fighters on all sides risk their lives for their cause. There are certainly many emotions motivating them, including patriotic feelings, desperation, and the righteousness of defending besieged lands or stopping terrorists. Among these emotions, the sweetness of imagined triumph might be the most powerful to lure people to the front.

The triumphant event functions as proof of the character's or actor's superiority. The event, the act, presented a conflict or challenge with a previously uncertain outcome, a test, contest, or temptation.[16] The more surprising the success of the protagonist, the more his prestige and power increase—and with it the co-experience by the recipient of the narration who feels elevated. Even if the success afterward does not consist in an actual increase of power, the feeling remains and rewards both the protagonist and the audience.

Astonishment as a Reward for Curiosity

Many stories and reports confront us with something unbelievable that we have never heard before and do not believe possible, but which is presented to us as true. We encounter such reports in the media, sometimes in the tone of the sensational, such as the story of Salvador Alvarenga, a fisherman who was lost at sea for more than a year until he was found stranded on the Marshall Islands in 2014. How did he survive on his simple raft? We also encounter remarkable stories in accounts from the natural world, such as astonishing observations regarding the behavior of a chimpanzee. It is not always a matter of natural phenomena; we sometimes hear stories from our friends that border on the unbelievable. The unbelievability may also stem from the fact that we know the person—we know, for example, that she happens to be shy and to lack self-confidence, and we simply do not believe that she could have delivered an improvised speech in front of a large audience when the situation was dire.

In former times people spoke of miracles that were connected with divine providence. More neutrally, one can say that the miracle is a singular phenomenon that defies direct repeatability or simulation.[17] This singularity of the miracle poses a challenge to the audience: Who or what caused it and why? How could it have come about? Should we believe it? There are no clear patterns or empirical data that allow us to embed the miracle in a known order, thereby explaining and normalizing it. The miraculousness of the miracle, in narrative terms, consists precisely in the fact that we mentally create various narratives to explain it, but these narratives simultaneously slip and fail to explain the event. Even the classification of the miracle as a "miracle"— that is, naming the state of exception (for instance, as divine intervention)—remains unsatisfactory and does not grasp the miracle. For even then it can be asked what caused this state of exception and who can recognize it. Is the divine intervention, for instance, legitimate? What motivated it?[18]

The likely narrative explanations slip away from the miracle, but this does not mean that we stop producing these versions of narratives aimed at explaining it. Wonder triggers an excess of narrative versions and possibilities. In fact, we may feel the marvel when a new narrative version and possibility opens like a secret door inviting a reassessment of all circumstances. Everything could be different. Even when one of these possibilities disappears—when it has turned out to be a false trail, a mere red herring—there can be a moment of wonder. As a recipient, one had "invested" in this variant of an explanation, imagined it, and now it disappears in one swoop, simply fizzles out. This fizzling away of the likely narrative explanations marks the cognitive structure of the miracle.

The hypothesis that I want to propose here adds to the narrative structure of miracles a dimension that is like the experience of emotion—namely, a reward for working on explanations that then, surprisingly, fail. This reward I call astonishment. We are astonished when we try to explain a miraculous event but find that the easy and likely explanations and narrative versions that could explain it disappear. Of course, searching for an explanation can lead to the positive "aha!" moment of understanding, but it can also lead to an insistent uncertainty. Narrative curiosity, the search for explanations, thus may lead to gains of knowledge but can also lead to the affect of astonishment. The fact that astonishment is also associated with cognitive gain, such as the reduction of some uncertainties or the insight that something is possible even if we do not know how, does not stand in the way of the spontaneous feeling of astonishment.[19]

In philosophical discussions about wonder, the moment of beginning has dominated since Plato and Aristotle. Amazement or wonder is the "drive towards the acquisition of knowledge," as Stefan Matuschek has summarized this figure of thought. Nevertheless, wonder in the form of astonishment can also be an end point. In the history of ideas, this end point is sometimes grasped as the limit of a taboo, for instance, in the Christian sense as seduction or also as the "absolute limit" of a near-impossible experience. For narrative curiosity, however, this possible end consists precisely in the fact that curiosity is exhausted in the beautiful feeling of amazement.[20]

If one takes up one of the standard models of emotions, the appraisal theory of emotions, then the narrative situation of wonder presents itself as follows: someone observes and evaluates (appraises) a certain situation as basically impossible or highly unlikely and then reacts with curiosity (produces narrative explanations, of which many fail).[21] The energy invested in curiosity can be transformed into astonishment when the puzzle without solution is acknowledged or when a highly unlikely explanation seems to emerge. The expected affect (astonishment) could thus already belong to the target motivations of curiosity that lead us toward this tricky situation. In that case, the situation is attractive to us because it is associated with and evokes the positive feeling of wonder and surprise.[22] Pursuing this idea shows that narrative curiosity is almost optimally aligned with this sense of wonder and symbiotically connected to it.

Astonishment, to borrow a turn of phrase from Lorraine Daston, is a "cognitive affect" that opens curiosity (what comes next) to wonder (how this can be explained) and then astonishment as a reward when the obvious explanations fail and vanish. This description of the narrative emotion of wonder (astonishment) gets some support from the fact that the feeling of wonder as an aesthetic emotion has a distinctly positive connotation. Wonder and astonishment thus qualify as a central motivation and goal of art.[23] We love art because—and when—we fail to explain it.

The narrative arc of wonder thus begins with a singular event that ignites curiosity. How, for instance, did the fisherman Alvarenga survive on a raft for more than a year on the ocean? Where did he get his drinking water, and how did he weather storms? Any quick explanation has to face improbability, so the curious person will consider and mentally test other possible explanations narratively. The very difficulty, perhaps the impossibility, of finding an explanation then awakens the sense of wonder as possible solutions are revealed and discarded. Each new explanation further increases the miraculous nature of the episode. The recipients reward themselves for the invention of numerous narrative variants, as if they were Salvador Alvarenga himself, who suddenly feels sand under his feet, staring at the land and unable to believe it.

Satisfaction with Deserved Punishment (Comeuppance)

In *Game of Thrones* when Queen Cersei Lannister instigates political murder and commits adultery, and when her son tortures his bride, we as readers or viewers expect punishment. At least, we count on some consequences—but for a long time, we are frustrated. Narration generates expectations, and without expectations there would be no narration.[24] Expectations include the possibility that they will not be met and they also always imply the delay of the end. In this respect, narration is the expectation of a delayed end. The end of a narrative is charged by these expectations and is under pressure to justify and legitimize the beginning with its expectations. Expectations of the delayed end may refer to the concrete occurrence of some facts, but likewise they are often morally charged with meaning: we hope for a good and deserved ending for the characters, and in some cases, the appropriate punishment.

Game of Thrones plays with its audience in this regard unlike most popular stories before it. In the beginning, the brave heroes, starting with Eddard Stark, vanish. They get executed or murdered. Redemption and comeuppance are on ice, postponed until a much-delayed future. However, revenge comes, as in the case of the Red Wedding, and it is certainly on the mind of the author, though he does not give it to his audience when they want it. When the TV series released the final episodes of the yet-to-be-published books, the audience was greatly disappointed: most people had bet on different characters and had hoped to see them elevated, instead of the sidekick who ends up taking the throne.

Expectations concern characters. Most likely, every narration includes at least one character who is assumed by the audience to have an inner life with planning, preferences, and feelings. Usually, the audience reacts to this character by hoping and wishing for an appropriate ending or outcome for him. The audience feels suspense, which concerns morality in a broad sense, insofar as suspense involves both a hoped-for and feared course of events that are each anticipated.[25] Accordingly, the audience of a narrative can and regularly does interpret the delayed ending as a reward or punishment for the protagonists. Narratives do not just happen to revolve around moral issues by chance; rather, they are moral by structure.

Narratives often end when the scores are settled. Debts are paid, deeds are fulfilled, and deserved outcomes emerge or fail to emerge. The good are rewarded, the bad are punished, and these outcomes elicit the feeling of satisfaction for the audience. This schema of expecting the morally deserved outcome is so strong that we not only associate it with fairy tales and religious texts, but it also characterizes the majority of works of world literature, as the literary scholar William Flesch argues.[26] In works of fiction, this expectation is so strong that we clearly react with indignation or disbelief when it is not fulfilled. Many modernist writers have used this indignation with great effect. We consider the outcome of a story to be unfair, undeserved if our favorite heroine has not been rewarded, or we resent the fact that a villain has escaped.

Although the term *morality* may sound old-fashioned to some people today and terms like *villain* may seem quaint or limited to superhero movies, moral sensibility is certainly not a feeling of the past. In many spheres today, we cultivate and refine a sensorium of what we consider right and wrong that in the past were not considered a domain of morality, including the new social media, environmental behavior, hierarchical work relationships. The heightened attention to morality in social relationships is noticeable, for example, in the many new expressions, such as *ghosting* or *microaggressions*, with which we expose and denounce transgressions.

Social psychologists and philosophers have argued whether morality and the sense of justice itself are closer to a feeling or whether they are more rational or rule-focused. Psychologists have defined a range of "moral foundations" (a term proposed by Jonathan Haidt) that guide behavior and that we sense strongly when they are violated. These foundations include fairness, care, authority, loyalty, and liberty. Still, people also follow reason and tend to agree on a wide range of ethical and moral questions.[27] For our investigation, the question is not whether specific moral foundations guide our decision making or reasonable ethical expectations. Rather, we can observe that narrative renderings of events very often culminate in either the feeling of moral satisfaction or its absence. Satisfaction is the pleasurable feeling when a delayed but hoped-for outcome of an action takes place. To be sure, the harmonious and moving feeling of satisfaction, on one hand, and the disappointment of its absence, on the other, are very unequal affective states. In the latter case, we do not tend to see a narrative as resolved.

If a narrative episode ends according to the "simple moral," the story has come to a good and expected end and is thus complete.[28] The emotional gain of that episode lies in the end of excitement and at the same time in the satisfaction that things have ended as they should. The feeling of satisfaction (comeuppance) is a distinctively narrative feeling, fed by the correspondence of the narrative arc's beginning and end. The bad deeds by the villain find their adequate conclusion in punishment. The moral tale thus offers us aesthetic symmetry, which indicates the extent to which those who think narratively map the arc of beginnings and endings onto one another. Similarly, the happy ending of a bullied and abused fairy-tale character like Cinderella also conforms to simple morality and the idea of a symmetrical beginning and end.

Narrative thinking, it should be emphasized here, does not consist of hurrying onward and onward. It is not linear. Rather, narrative thinking comprises an organizing mind that attempts to relate the diverse elements of the story to each other and join them together. Nowhere does this succeed as clearly as it does in the moral end and its associated feeling of satisfaction, because there the end is evaluated as the exact equivalent of the beginning.

In the first chapter we asked what constitutes a narrative episode, recalling the nineteenth-century literary theory of Gustav Freytag: a narrative sequence is an episode that portrays a turnover of active and passive experiences in the main character. The moral narrative episode conforms to this idea. The character's actions (active) are rewarded or punished at the end by the narrative's resolution (passive).

The feeling of satisfaction, which is the issue here, is the feeling of the release of a tension; it also provides the sense of aesthetic harmony. With the first moral or immoral acts, the expectation of just deserts is generated. In the feeling of satisfaction, this delayed expectation comes to fruition. Now the audience can relax and move on, forgetting the first actions since they have been adequately compensated. Something similar applies not only to morality in the narrow sense, but also to many other expectations—for example, we also expect hard work to be rewarded. The readers of this book may breathe a sigh of relief at the end of each chapter as a reward for the hard work of reading it; I, too, will at some point come to an end.

William Flesch suggests that this stress on morality in works of fiction spurs us on to behave morally in everyday life, to avoid punishment, and to engage in "altruistic punishment."[29] Altruistic punishment emphasizes the tendency of punishers to act for the good of the community rather than their own self-interest, as they themselves take on risks, such as revenge from those being punished; altruistic punishment helps the community, as it sends a signal that morality needs to be upheld and thus deters others from bad actions. It should be noted, however, that this punishing is often not completely selfless and altruistic: punishers bring about the compensation they deem appropriate and accordingly receive the emotional reward of satisfaction.

Of course, not all episodes have a good ending, both in fiction and in real life, and some villains get away with their wrongdoing. But that does not mean that morality does not also structure our expectations in these narratives. Even and especially in cases where things do not go as we hope, we maintain the balance in our mind that something is not yet resolved. The feeling remains as the hope that everything could turn out differently in some undetermined future and that there will be a final verdict. Morality persists in this very expectation that the protagonists deserved better, for instance. Even in the case of long-delayed expectations of moral satisfaction, there is some emotional payoff in our knowledge that things should turn out differently, but usually this knowledge is not perceived as being as aesthetically pleasing as satisfaction.

There are also, of course, many cases in which the hoped-for reward or punishment fails to materialize at the end. This again applies to both real life and fiction. We may judge narratives that lack satisfaction as bad, unfair, or stupid. This judgment can result in a variety of feelings of dissatisfaction, anger, or disappointment directed at the author of the story, specific characters in the plot, the storyteller, or our belief in humanity as a whole. For example, a narrative that evokes pity for a wrongdoer because of his or her harsh childhood undermines an emotionally satisfying ending, because neither punishment nor reward seems appropriate; this pity leads to doubt and despair about humanity itself.[30]

The absence of reward or punishment, however, can also be discharged as anger, outrage, and agitation. These emotional reactions share an index for the future. In this sense, Amartya Sen considers injustice experienced by an individual to be the origin of the feeling of justice—and this includes the search for the still missing future satisfaction and thus for satisfaction generally.[31] Injustices expressed in true and fictional stories can inspire revolutions and cause social movements, leading nationalists, climate activists, and feminists to take to the streets. The Arab Spring in the early 2010s brought down governments in Tunisia, Egypt, and Libya. We hear a narrative episode that stirs us morally but does not give us an appropriate ending, and we feel called to act ourselves to bring about that better ending.

Those who are so stirred demonstrate that the narrative is not over but instead requires another, future ending that will provide satisfaction and gratification. Here again we see how clearly narration and our own actions are interwoven, for in these cases we become part of the story that we may have just been told. The narration spills over onto us and sweeps us along in its narrative wave.

Finally, there are narratives that lack a morally good and appropriate ending— and consequently satisfaction—but still achieve a goal aesthetically and invoke other emotions. This is the case, for example, in *Romeo and Juliet*, as well as other tragic, sentimental, or nostalgic stories. In these stories, we notice the sad injustice of the course of the narration, which, however, takes a compelling and thus tragic (because it cannot be changed) course. We may find ourselves emotionally in a back and forth between the work as it is and our belief that it should all be different, our desire that it all be different. Intervention is not possible in works of art, so the despair and the hope of satisfaction are in a dynamic relationship. Apparently, we have learned to appreciate this dynamic stasis as aesthetically positive.[32] Perhaps it is even the case that the compassion the audience feels for the suffering of Juliet (and Romeo) includes a positive feeling representing a kind of substitute reward for the lack of a positive outcome insofar as the audience members become aware of their involvement. The moving feeling of compassion for an innocent person so close to rescue provides a small dose of satisfaction by showing that one is rooting for the right side.

Being-Moved, Especially as a Result of Recognition

"Being-moved" is a central narrative emotion that has also been put forward by Winfried Menninghaus and his colleagues as a possible candidate for the basic form of all aesthetic emotions.[33] So what moves us? What are the elements of being moved?

A starting assumption can be that the emotion of being-moved has to do with temporal processes and, more concretely, with narrative structures. A painting itself can also be touching, but it seems reasonable to assume we are being moved by the story in the painting. At least, non-representational art is rarely described as moving. What would serve as a good example of being moved by a narrative?

In surveys among my students over the years, one story was regularly named the most moving and touching: "Cinderella." The surveys are certainly not representative, but they give us a hint, especially since "Cinderella" is indeed one of the most popular fairy tales in the world.[34] What makes this story so touching? "Cinderella" is, in many ways, a clear case of evoking pity with a stereotypical gender portrayal. The death of the weak father, the unjust stepmother, and the wickedness of the sisters combined with Cinderella's altruism and compassion clearly mark her as a vulnerable and therefore good character. One of the strongest triggers of empathy (by the audience) lies in observing a person who has empathy herself; Cinderella, for example, helps little animals and treats them with respect.[35] Earlier we touched upon the peculiar role of vulnerability in the Grimms' folk fairy tales. "Cinderella" has all of these characteristics in common with most other fairy tales. Therefore, these features alone probably cannot explain its extraordinary success.

"Cinderella" adds an important feature to the listed elements of other fairy tales. Cinderella is a double character. On one hand, she is the lowly maid, and on the other, the beautiful stranger of the ball. In neither of these roles is she fully recognized. She is not appreciated for the work she does as a maid, and she does not get recognized for who she is at the ball. The act of recognition and displaying herself becomes dramatized. She is allowed to show herself in her beauty by means of magical help. In general, showing one's true self presents a powerful moment, combining a sense of presence with empathy and identification. In "Cinderella" the moment of self-display and recognition is delayed as she, the maid, cannot be recognized and identified as the beautiful stranger. Only the magical shoe can link her two appearances. The moving culmination of the story then lies in discovering the beautiful stranger in the lowly maid. At the heart of the tale stands a scene of recognition: Cinderella is recognized by the prince, the malicious stepmother, and the whole kingdom.

First we must ask: what does showing oneself and being recognized have to do with the emotion of being-moved? We should emphasize that the beauty of Cinderella cannot, of course, be shown visually in a narrative, but must be imagined. Moreover, it is not unimportant that the readers already know who Cinderella is before the official reveal.

Now we are equipped to consider recognition. In *Poetics*, Aristotle suggested that recognition (*anagnorisis*) along with *peripeteia* (the turning point) is one of the two central structural moments of tragedy that make it emotionally powerful.[36] Scholars have discussed the implications of anagnorisis in Aristotle and tend to fall into two camps. The first camp focuses on the intellectual work of understanding that takes place in recognition.[37] The second grasps its affective-emotional side, in which recognition becomes an event. In the following, we will deal with this second interpretation.

The first question we can ask is this: Who is actually doing the work of recognition? It should be noted that it is unclear whose perspective the audience takes in a recognition scene—the perspective of the recognizers or of the recognized? When Cinderella reveals herself and slips into the shoe, do we see this event through her eyes, through those of the prince, or even through those of the envious stepsisters? The same question can also be asked in other famous scenes of recognition—for example, the return of Odysseus in the twenty-second canto of the *Odyssey*, when he reveals himself to be who he is—namely, the enemy of the suitors who aim for the hand of his wife. In that moment, the old man they had formerly mocked has just bested them in archery and suddenly regained his strong appearance. He yells out at them, bow in hand: "Ye dogs, ye thought that I should never more come home from the land of the Trojans, seeing that ye wasted my house, and lay with the maidservants by force. . . . now over you one and all have the cords of destruction been made fast."[38] In the edition of my childhood, this speech was simplified and shortened, if my memory does not deceive me: "Look into my eyes. I am the destroyer of Troy." At this point, the audience can take the perspective of Odysseus in his true role as he declares his triumphant homecoming, now finally unloading, arrow by arrow, his long-saved rage after all his humiliations. Or they can hear this call from the point of view of the horrified suitors, who suddenly find themselves the target of a mythical war hero whom they considered long dead. The position or role of the listeners is not predestined, at least not in terms of either position. This is precisely the central property of recognition, as I would like to suggest: as audience, we witness and co-experience it from *both* positions at once.

Here is my proposition: in a scene of recognition, the two positions of recognizer and recognized are intimately intertwined, and the gaze of the audience oscillates. The audience is drawn to both perspectives and is thus tossed back and forth from one perspective to the other. The one who is recognized (that is, Cinderella or Odysseus) sees in the other precisely the gaze of recognition and the effect this recognition has. The one who recognizes (the prince and the stepsisters or the suitors in the *Odyssey*) not only recognizes the other but at the same time sees how the recognized other observes the arrival of this recognition. Cinderella sees how the prince perceives her as a princess again; Odysseus registers how the suitors freeze in recognition. It is in this way that the audience experiences being "seen" from each perspective. Odysseus experiences how he is seen from the eyes of the suitor Antinous, and so forth. This means the audience experiences the effect of both seeing and being seen, back and forth, at the same time. Odysseus "sees" in the suitor Antinous how the latter, in turn, recognizes him; and Antinous "sees" how Odysseus "sees" the suitor's fright as the effect of recognition. That is to say, the one perspective accesses the other, integrates them, so that after the first entrance the distinction of which shoes the spectator has slipped into is lost. In "Cinderella" as well as in the *Odyssey*, there are three perspectives available: the recognized hero, the happy comrades, and the punished third parties (that is, Cinderella, the prince, and the stepmother). The audience behind the fourth wall could function as additional perspective.

Recognition (anagnorisis) consists therefore not just of a simple, one-time transformation of ignorance into knowledge, as Aristotle so wonderfully formulates, but a permanent back and forth, an oscillation between two or more positions, an *event of recognition*. In the back and forth, the audience members experience themselves as recognizers and as recognized at the same time. Nevertheless, the audience never quite feels directly as one with either; it recognizes and feels how the recognized is recognized by its counterpart and in turn recognizes itself in this perspective. The audience thus experiences how it recognizes being recognized as recognized . . .

What I want to hint at here is that this form of observation does not come to a conclusion but increases continuously. In this event or experience of recognition, the audience begins mental motion. Indeed, aesthetic theories have long used the vocabulary of movement for this process.[39] Aristotle talks about the shaking in peripeteia and anagnorisis. Immanuel Kant translates the feeling of the sublime as an oscillation that goes back and forth between concept and sensory perception. And perhaps this very movement is a part of what we call "being moved" in the sense of the Latin *movere*.

The observed event becomes an event of observation. The observed event of a revelation or the uncovering of a person's identity becomes an event for the audience, but *not* because its members are surprised like the other parties. After all, the audience for the most canonical scenes of anagnorisis is already in the know, aware of the identity of the parties (Odysseus, Cinderella, Iphigenia) from the beginning. The event of the revelation itself is thus nonexistent for the audience. In fact, it is probably helpful for the oscillation if the audience members already know who they are dealing with and for which revelation they have to wait. That way, they don't have to focus on information and knowledge. Rather, the event for the audience consists in this movement of perspectives. As soon as they engage with one perspective, that perspective slips away and they are drawn into another. Recognition doubles the event, pulling it from the level of motif into a performative process of back-and-forth observation.

"Cinderella," then, according to the reasoning in these remarks, is moving because we, as an audience, participate in the act of recognizing ourselves and being recognized. In fact, we participate in a highly active way. Our imagination swings back and forth between recognizing and being recognized (positively by the prince or negatively by the stepsisters); we also recognize the effect of recognition on others. This can bring tears to one's eyes because this moment, which never stops but continues its back and forth, is a reward for the torments of suffering before. However strong this torment may have been, it is now compensated by this moment. Her act of appearing, showing herself, is permanent and merges with her character. She is the one who appears. In the sense described above, her beauty, not in her physical appearance, *moves* us. In the same way, the revenge and retribution of Odysseus shake us.

The literary scholar Eva Geulen, building on work by Clemens Lugowski, has highlighted another dimension of this being-moved that also exhibits a structure of anagnorisis: the recognition of oneself in retrospect. Although recognition in the examples above had the function of uniting two sides of the self (Cinderella as maid and unknown stranger, Odysseus as old man and destroyer of Troy), Geulen emphasizes a divisive dimension of this self-recognition: "In 19th-century narrative literature, one typically recognizes oneself in what one has and has not become." In recognizing oneself in retrospect, the life not lived is recognized and evoked again. This self-recognition is moving, according to Franco Moretti, because in this recognition of unlived life lies both an acceptance and a farewell to the past, a subordination to the principle of reality that nostalgically conjures up the lost as lost. In this respect, there is an "affirmative trait" within being-moved, according to Geulen, that becomes "suspicious" from the point of view of modern literature or critical literary studies.[40]

For the present investigation, we should emphasize the positive dimension of the back and forth, the dialectic of appropriation of the past and the farewell, and the recognition of that which has never been. Even the memory of children who died at an early age could become moving. This could be due to the affirmation of the lost as lost, as Geulen suggests with Moretti. It could also be due to the oscillation of comparing the present life with the not-lived life that could have been.

Perhaps this dimension of self-recognition also plays a role for Cinderella: her happy life at the end is the non-lived life by the audience.[41] In this sense, then, being moved is the reward for participating in recognition. We are moved because we as audience become caught up in the act of appearing and being recognized.

Is being-moved, then, a feeling like other feelings or even an emotion? This indeed can be argued.[42] For our context, one could also say that being-moved is only indirectly a feeling or an emotion and only becomes an emotion insofar as it arises from a feedback-loop of being observed. Because this feedback loop never ends, it is understood as meaningful, deep, important, and thus akin to an emotion.

Even if some of my readers take issue with the complexity of the previous considerations concerning the paradoxical oscillation of anagnorisis, here is a proposal for a minimal consensus: narratives can move us and shake us to the core. Being moved can be an effect of a narrative, and thus the expectation or hope of being moved can serve as stimulus for narrative thinking. It also can be a reward, as it stirs us with life and feelings of humanity. Being-moved thus can be seen as a powerful narrative emotion.

Finally, it should be noted that recognition (our core case for being-moved) probably has a stronger presence in works of fiction than in real life, but it can appear there as well. One of my recently deceased mother's happiest memories comes from her earliest youth. Her father, my grandfather, was first missing in action during World War II and then held as a prisoner of war. For a long time there had been no sign of life from him. Many prisoners of war vanished in those years. My mother tells how, as a child in Hamburg, she was sitting at the window of a relative's house that had not been destroyed by the air raids, and she saw a stranger in the distance—and suddenly she heard a whistling sound that she had not heard in years, which sent her into a shock of joy. It was her father returning home unexpectedly.

The narration of this little episode has always brought tears to my eyes. Even now I cannot write it down without emotion and tears.[43] It is the *narrative* rendering of this episode that turns the homecoming and the recognition into an event in which others, like me, can participate and be moved even after seventy-five years.

Surprise and Novelty

Narratives are fixated on events. Without at least the possibility and expectation that something will happen, there is likely no narrative and storytelling. Events are incisive. They are, in the words of the narratologist Wolf Schmid, relevant, unpredictable, persistent, and irreversible.[44] The situations before and after the event are fundamentally and irreversibly different, at least in some respects. This is obviously true for unpredictable catastrophes and the drastic adventures in novels, but it is also true to a certain extent for everyday events: the long-shot application for a job turns out unexpectedly well, and with a new job a new world opens up; the spontaneous invitation to lunch throws one into a new social group; giving up an apartment uproots a person; or a chance encounter in the subway shakes one out of a deep melancholy. What these events have in common is the abrupt confrontation of mundane everyday life with something new. It is true for every event that the worlds before and after the event cannot be reconciled. Slavoj Žižek has expressed this idea by calling an event "a change in the very frame through which we perceive the world and engage in it."[45] In other words, one can predict neither the event nor the state resulting from it before the event occurs. Goethe formulated the following at the beginning of his novel *Wilhelm Meister's Apprenticeship and Travel* (1795), the novel that has become the model for the formation novel or coming-of-age tale that has dominated novels ever since: "It is so pleasant to think, with composure and satisfaction, of many obstacles, which often with painful feelings we may have regarded as invincible,—pleasant to compare what we now are with what we then were struggling to become."[46]

According to this reflection, the states before and after an event are not symmetrically opposed to each other; rather, the peculiarity of the before consists in considering the after to be impossible. Conversely, from the state of the after, it is at least a delightfully paradoxical (and in Goethe's world thus "pleasant") task to consider one's earlier point of view of deeming the after impossible. The break between the two states, the interruption by the events, is complete.

Of course, we do know that events take place despite all these paradoxes. They are simply unpredictable. If we follow Goethe and Žižek, events are less concerned with physical occurrences and anomalies, such as an earthquake leveling a city, and more concerned with shattering the expectations of those affected. That is, events surprise. Surprise is an integral part of all narratives in that every event is surprising to some degree. Surprise is apparently not an incidental aspect of narratives. In the retelling studies discussed earlier, we saw that the degree of surprise in serial-reproduction tasks is passed on very precisely and thus is maintained while many other facts deteriorate. In fact, the degree of surprise was apparently so important to the narrators that it was at least partially preserved even when the original surprising event was omitted or completely changed in the retelling. In these cases, the surprise was moved to other, previously hardly surprising events.[47]

Is surprise a reward for narrative thinking and does it qualify as a narrative emotion? In the case of most emotions, we usually know exactly whether we like the emotion or not and whether it is positive or negative. It is a bit less clear whether surprise should be considered negative or positive. One possibility is that surprises are seen as negative at the beginning because expectations are contradicted, but they result in a positive state at the end when people have integrated the events and information into a new coherent structure.[48] In this respect, traumatic events would remain negative because they are unfinished, while other surprising events that culminate in a new whole might fit with Goethe's paradoxical feeling of being pleasant. The integrating work of the surprise could render it positive in the end.

Narrative surprises might offer a reward in another respect. Surprises present something new. People tend to evaluate stagnation as negative overall, whereas novelty is positive, even if this novelty is not accompanied by other rewards or improvements.[49] Edgar Dubourg and Nicolas Baumard go so far as to explain this tendential preference for novelty, especially in geographic terms of exploration, as the central evolutionary-adaptive incentive of fiction that, like Tolkien's *Lord of the Rings*, allows us to explore a new world.[50] I am not fully convinced by this argument, since imaginary fiction, such as works by Tolkien, J. K. Rowling, Liu Cixin, or George R. R. Martin, is a relatively recent phenomenon, and similar epics of the past, such as Homer's *Odyssey*, are relatively few. The appeal of these works, I believe, may instead consist in the re-enchantment of the modern, disenchanted world. These works also have another feature in common—namely, the figure of the "chosen one" as hero who can save the world, and this figure is especially attractive to our modern state where people feel the loss of individuality.

Thus, instead of looking for the value of novelty in imaginary fiction in fantastic geography, it seems to me more fitting to locate it in the event structure. Surprise provides the exact measure of a novelty that we enjoy: if there is too much novelty and too many events in a story, it results in chaos and is no longer considered a surprise. This applies to works of fiction as well as to life stories. If someone gives me insights into his life and these show an excess of information and confusion, I find it difficult to evaluate each event as a surprise and to follow the story. My feelings and consequently my empathy (understood again as co-experiencing the situation of another) are no longer evoked. Conversely, many people's tales of their life suffer from uneventfulness. Only exceptionally great narrators, such as Adalbert Stifter, succeed in charging even relatively uneventful narratives with positive tension. The middle ground is the story where a surprising event splits a before and after. Here we find a perfect puzzle, a shock, and a chance for a pleasant reintegration that is everything but boring. In short, while surprises in narratives are not always positive or automatically rewarding, the unexpected can offer the reward of movement, novelty, intellectual titillation, and delight. Each surprise shakes us out of the valley of uniformity and carries hope for the new that we do not dare to express.

Laughter as Diffusion of Embarrassment

There are many scenes from my life that I don't like to remember. These include, above all, many embarrassing moments in which things didn't go the way I wanted and I didn't cut a good figure in front of my fellow human beings. There are episodes from my childhood that even now I do not tell because they are still connected with shame and embarrassment. Did I really . . . ? I hesitate to think of it.

Embarrassment and shame clearly have a repressive function, contributing to the oppression of individuals and groups. They are condensed in taboo and stigma.[51] Stigma sticks and is transmitted across generations. Even in the retelling experiments reported above, embarrassment was clearly preserved and passed down through numerous iterations.

I've been told that my grandmother, a young German aristocrat at the time, was presented to the Empress Mother at a ball during World War I. The German Empire celebrated its own version of pomp and circumstance at that time, and apparently my grandmother made a mistake in etiquette and addressed the mother of William II with the wrong title, or something like that. This was so embarrassing to her that she swore never to go to the court again. From then on, she completely avoided the nobility and instead studied at Heidelberg, working as an assistant to a renowned philosopher. One episode of embarrassment changed her life. This too is a way to academia and philosophy.

Another arena of great embarrassment is, of course, to be found in dating. My youth is full of memories of how I made a fool of myself. When I arrived in America to study for a Fulbright year and introduced myself at the registrar's office in Baltimore, there was exactly one other student there. It was still two weeks until the semester started and hardly anyone was on campus. I was eager to meet other students and asked my fellow student if she would like to meet up in the evening. She agreed and we arranged for me to pick her up. When I arrived there in the evening and she opened the door, I knew immediately that something had gone wrong. She had put on a fancy evening dress and makeup—apparently she was expecting a date. I, on the other hand, with my low-key European manners, raised in an era past pomp and circumstance, had cycled over, thinking we were just going to a university pub. Europe back then simply did not have a dating code. I had probably expressed myself stupidly. My cultural misunderstanding was obviously embarrassing to me but more so for her, and we had to make the best of the evening. (No, we did not get married.) The idea of calling a cab and getting a ride to a fancy restaurant didn't occur to me at the time.

Ritualized forms of dating manage expectations by coordinating the behavior of those who master the code. The standard result is to reduce errors, which means that cultural misunderstandings like mine or class divisions like my grandmother's stand out even more. Standing out means embarrassment. Certainly, the tremendous success of dating apps in the past twenty years is due in part to the avoidance of shame. Anyone who clicks on Tinder at least knows that the others are in principle also looking for a date. This avoids the faux pas in advance.

The reason why shame and embarrassment (sometimes referred to as *self-conscious emotions*) appear in this chapter is the possibility for their transformation.[52] Embarrassment can disgrace us and, even worse, drive us to philosophy. However, we may also be able to develop strategies to deal with embarrassment. One of them is that we can laugh about it too. Usually, it is the others who laugh. However, we can also tell our own story about an embarrassing scene—and then laugh along with others about ourselves. We may even tell the story precisely for the purpose of bringing about this effect.

As a doctoral student I spent two years in Berlin, where I made my money teaching German in a small language school for international professionals. The small group of learners included members of high society, and so we were all invited to dinner by the American consul general, as his wife was in my class. At the door, in front of my students, I vigorously shook hands with the gentleman in the tailcoat and expressed my joy in meeting him—until he whispered to me that he was the butler (maybe my grandmother was smiling at me then). I can recount this cheerfully now, but at that moment I certainly felt the embarrassment of the episode.

Embarrassing episodes can become highly narratable for several reasons. One is that they involve a dramatic switch in perspective. The person talking about her experience suddenly seems to become her own audience, laughing about what she did. This also involves a switch from active involvement to passive experience in the scene, as the speaker presents herself as the one who committed the embarrassing act but also as the person who endured the public humiliation, and now—in the retelling—as the narrator who has the wisdom gained from the experience to understand its humor. Finally, embarrassment often goes hand in hand with surprise, another clear narrative emotion.

The potential for stigma and exclusion is clearly always evoked with scenes of embarrassment, but laughing along is a strategy for destigmatization. When I laugh along, I show that I am part of the in-group that makes fun of the out-group. By appropriating the episode of my own embarrassment as a narrator, I transform the stigmatizing event into something that moves others and me to laughter. In this way, Charles Baudelaire depicted the proud man who believes himself in control until he trips over a simple stone; only the philosopher succeeds in laughing about himself.[53] Humor here becomes a narrative affect that defuses a dicey situation and at the same time brings it to a close. Instead of ongoing stigma, the episode can end in laughter.

This positive power of laughter has been discovered and cultivated for millennia in comedies and, more recently, in sitcoms. Although not every effect of stigmatization can be reversed in this way, the therapeutic effect of humor should not be underestimated. It is, as I will at least suggest as an assumption, definitely in competition with other forms of individual therapy. Those who arrange their lives in humorous narrative episodes are both liberated and rewarded.

Love and Eroticism as Narrative Emotions

If we include love among the narrative emotions here, a few objections must be cleared up in advance. These include the question of whether love is best described as an emotion or as something else, such as a state. Second, there is the question of what kind of emotion it might be. Is it an observer-focused narrative emotion, or is it "only" an empathetic-narrative emotion, because love is evoked in recipients only insofar as they identify with characters in the narrative and thus simply experience love. A friend of mine, an eminent cultural critic, once revealed to me that he had fallen head over heels in love with Neytiri from the film *Avatar* (2009).

The answer to the first question is simple, insofar as it is in part simply a matter of conceptual and definitional demarcations. Love not an emotion? Try to explain this to a layperson. However, for a long time academic emotion research in psychology absurdly did not consider love a "basic emotion" precisely because it proceeds like a "plot"; it is thus like a narrative and might also last a long time.[54] Love just didn't want to conform to the concept psychologists had created according to which emotions are short-lived. Now it is widely recognized as an actual and central emotion. Love, because of its long time intervals, has sometimes been classified as a state and therefore not an emotion. However, even lifelong, stable forms of relationships, such as love of God or love for one's own child, are not constant and unchanging—they develop and are experienced concretely and with differences at different moments in time. Contemporary psychology therefore also emphasizes the temporal moment in love and now values love as an emotion: love is the "pleasant and momentary experience of a connection with another person (or persons)" or a "momentary psychological phenomenon that is coexperienced by any two or interacting people."[55] Love is felt one moment at a time, and these moments follow from previous moments and sequences. This suggests that narratives, with their emphasis on one moment at a time, coupled with narrative progression, can at least approach love.[56]

The answer to the second question leads us to a more precise determination of what love is as a narrative emotion. What is the narrative arc of love and love stories? And what does the participation of the audience consist of? Concretely, does the audience also fall in love? Less technically, what happens when we hear love stories? Why should we care about the love and sex adventures of others?

It might seem obvious here to distinguish between love and sex, or between erotic texts, in which the act itself is foregrounded, and romantic love stories. At the same time, however, there is a long and popular literary tradition in which romantic love and sexual consummation come together: the story of the loving couple overcoming all obstacles and coming together ends in church or in bed. It is this narrative of love and sex that I will explore here.

This means that erotic tales and even pornography also belong in this context, because hardly any pornographic representation can do without narration. Here, too, in the sense of typical narrative patterns, the story is about overcoming obstacles—even if these are only to be found in the clothes that get in the way. And as soon as the emotional reward is achieved in these tales, everything is really over.[57]

Nevertheless, there is an interesting difference between everyday narratives and works of fiction. When a friend tells me about his new infatuation, I am correspondingly moved. But do I identify with my friend? Yes and no. If I identify too much with him, would I now be in love with the same partner? This would make us rivals.[58] Here we find a hint that there is another dimension to the empathetic component of co-experiencing another's situation. A first and very general assumption is that the way we tell love shapes love.[59] The question is how.

Michel Foucault has argued along these lines that there has been a historical turn in Western cultures since antiquity in which the act of sex is substituted by discourses about it. For Foucault, such substitution typically leads him to considerations of power: what is substituted by speech and discourse is thereby also controlled, regulated, and thus dominated by something else, namely, discourse. Foucault substantiates this hypothesis, among other things, by means of many interesting case studies that suggest an increasing interest in the verbalization and differentiation of erotic acts.[60] In fact, Foucault's methodological optics of focusing on such discourses that control behavior consequently only allow him to see control, power, and policing. What I would like to suggest here is that such discourses and verbalization also indicate other possibilities that are less concerned with control, namely, narrative developments that can resolve and transform tensions: romantic love, physical love, and sex can be narrated. This leads us back to our question: How and for what reason are erotic relationships narrated?

Let's start with the space of the fictional. The basic model of the novel in the Occident and in the tradition of antiquity over many centuries can be described in the simple "scheme of finding each other, separation, and reunion of a loving or married couple."[61] This basic scheme of the novel is so powerful that we might wonder if it eclipses the (male-focused) archetypal scheme proposed by Joseph Campbell's *The Hero with a Thousand Faces* (1949), in which the hero is a lone wolf who has some adventures and then returns enriched.[62] Two narrative structural elements stand out in the love novel and romance stories: the overcoming of resistance after a separation and the reunion of the lovers. In the great love stories of Western literature, this resistance between the lovers regularly takes the form of a social obstacle. Cupid and Psyche must prevail against jealousy. Romeo and Juliet despair about their families. Cinderella must outwit her stepmother, and numerous great love stories show the lack of propriety when princes or princesses fall in love not with nobles but with commoners. In the nineteenth century, the scheme persists, but now the resistance to be overcome is increasingly situated within the characters themselves, who must overcome their stubbornness and pride, like Jane Austen's Elizabeth Bennet and Mr. Darcy. Even the seduction stories of a Casanova have a similar form. Casanova has a parasitic strategy of recommending himself to his chosen one. He often finds her in a tense situation, for example, when she is being harassed by an aggressive lover or oppressed by a hostile institution, or he makes her believe that she is being oppressed. In this situation, he can then offer himself as a confidant and earn her trust. Her gratitude for his apparent selflessness then becomes a trap. The confidant of the woman suddenly becomes a self-serving lover. But even in this case, and despite Casanova's cunning, the narrative pattern follows the same large pattern of separation and reunion. In his reports, Casanova usually assigns resistance to an external force, since he describes his chosen ones as willing lovers, and refocuses resistance on the former, jealous lover or an adverse institution like the church, which the two lovers now have to outwit.[63]

In what we have considered so far, the audience is concerned with the lovers' goal, namely, the reunion. This goal might offer some suspense and satisfaction but it is in itself not specifically romantic or bound to love.

We get further if we consider the specific form of union or reunion in erotic tales. Not just any resistances are eliminated here, but specifically those standing in the way of an erotic coupling. From a narrative point of view, what's special about such erotic and romantic coupling? First of all, it involves a physical reaction for the audience with tingling skin and a faster pulse. Witnessing unions of love never leaves us cold. Our entire physiology, including that of smell and touch, is addressed.[64] Furthermore, the coupling also includes changes in the narrative characters' status: they cease to be what they were before.

From now on they appear as a fused couple, and the persons they were before are a matter of the past. In a narrative sense, we can say that the characters have fulfilled their potential. The characters have achieved their destiny and are now being absorbed into a new entity. This rite of passage is irreversible. In narrative terms, therefore, the loving union marks a clear end. Accordingly, many of the great romantic tales end just here, with the happy ending. The audience can now withdraw from the story or establish a new relationship with the changed characters.

Often this dissolution of the character in the erotic fusion applies to one of the characters more than the other. A powerful tradition presents especially the life of the female characters as ending with marriage, even though further adventures may await the man. For the woman, however, few future roles are available. The character of the "fallen woman," with its massively repressive power over centuries, also belongs in this context, insofar as it contains the idea that the one-time sexual act outside of marriage marks and shapes the woman for the rest of her life. Again, this scheme is typically applied to women who get seduced. The motif of seduction also exists for male protagonists, such as in the plot scheme where a woman aims to catch a desirable man, as in the Marilyn Monroe film *How to Marry a Millionaire* (1953), but the pattern here does not include the sexist degradation of the male.

With the happy ending and the erotic union we come to the observer-focused emotion of love: the culmination of the tale, the end of suspense, and the physical excitement go hand in hand with the dissolution of the contours of the narrative character. Love stories cast the person or character in the role of a potential lover or object of desire. With the end of the story, with the lovers' union, this role ceases to exist: the fulfillment of the promise of the characters' potential is both their climax and their end. The dissolution of the character throws the audience members back to themselves and out of the narrative. As audience, we are freed from the lovers and can leave them alone again.

The arc of the narrative emotion of love slides over from the excitement of the characters (empathy-focus) into the real life of the audience (observer-focus). The narrative ends with a happy and stimulating emotion, and this emotion and excitement remain when we as audience come back to ourselves. The empathetic dimension of co-experiencing (love, desire, excitement) and the observer-focused emotions (suspense, dissolution as fulfillment) can thus enhance each other. It is no wonder that love stories are among the most popular narratives in the world.

What Have We Learned?

Emotions and feelings have a number of important functions that contribute to our survival in evolutionary terms. These include an expressive, communicative dimension: like early humans in the steppe, today's humans show others what they need via displays of emotions. Communicating emotions is powerful, as we are wired to react to emotions. For many expressed emotions that we observe in others, we match these emotions; that means our brains utilize the same or similar routines for having and observing that emotion. This is often referred to as the "perception action model of empathy" where perception (observation of an emotion like pain) and action (having that emotion oneself) are linked.[65] A common effect is that the display of emotions causes us to do something. Those who cry are helped. Whoever expresses love is more likely to find partners.

Inner feelings also have an important function for us by guiding and directing us to what we should do: should we approach others or avoid them? Such guiding feelings are faster and more effective than strenuous rational calculations would be.[66]

The hypothesis of this chapter was that there is a group of distinct *narrative emotions* that reward narrative thinking. Each of these emotions does this in a different way, but what they have in common is that they spur on and motivate narrative thinking. Narrative thinking offers numerous survival benefits, expressed in better remembering, planning, reacting, and orienting, and potentially in overcoming depression and trauma. Narratives and rendering one's own life as a heroic story of overcoming obstacles increase perceptions of meaningfulness and thus emotional health. Moreover, if we add a collective dimension to narrative, enhanced group formation (bonding, cohesion, and coordination) and information and experience sharing (learning) come into play. We can posit, therefore, that narrative emotions make substantial contributions to our survival.[67]

There are three aspects of this hypothesis that narrative emotions reward narrative thinking that we should emphasize. First, the expected rewarding emotion draws us to engage with an episode in the first place. To do this, of course, the emotional incentives must first be associated with specific sequences and thus acquired and practiced. This means that the specific narrative sequences have to be learned and can accordingly turn out very differently in different cultures. Not every narrative ending, such as the bittersweet feeling at the end of a melodrama or nostalgia, makes sense to a child or a migrant from another culture.

Second, a number of the reward processes described show how emotions are *structurally* linked to the end of a narrative. For instance, when morally good people are rewarded at the end, this feels gratifying and proper (satisfaction, justice). The beginning (that is, the good deeds and dispositions of these people) coincides with the (rewarding) end. This structure of a matching beginning and end confirms our hopes and expectations. In such cases, then, the rewarding emotions at the end spring from a cognitively recognized narrative sequence that combines the episode in retrospect. We linked the match of beginning and end with aesthetic emotions related to symmetry. This effect suggests that better narrative skills and more accurate narrative memory lead to clearer episode boundaries (where something comes to an end) and clearer emotions. Similar considerations could be made for several of the narrative emotions discussed above where beginning and end do not stand in symmetry but nevertheless have some clear relationship and where the recipient of the narrative will engage in some retrospective construction to connect the two. A case in point is surprise, where hindsight knowledge leads to a revision of one's earlier knowledge, suspense, and emotional response.

Third, the audience is the agent of emotion in the narrative. This aspect should be clear, but it must be emphasized again to avoid confusion between characters and audience. After all, the characters usually do not know that they are part of a narrative process. Narrative thinking happens in the mind and brain of the audience.

Narrative emotions motivate us to engage in stories, teach us about structures of narratives, and turn the audience into the secret agent of the stories. In exploring our hypothesis that narrative emotions reward narrative thinking, we have looked at several emotions—triumph, astonishment, satisfaction, being-moved, surprise, and love—but the list of additional narrative conditions is long. We have not touched on the sublime feelings in tragic stories or the feeling of redemption in particularly exciting stories, for example. (We have also not yet focused on suspense, because suspense itself is probably not the reward of narrative thinking; if it is, it is only insofar as it pulls the listener out of lethargy and boredom. We will come to the special role of suspense later.)

A number of further investigations can significantly broaden the spectrum discussed so far. Christiane Voss, in her book *Narrative Emotions*, has already made an attempt to anchor emotions as a central object of literary studies.[68] Sianne Ngai has proposed an idea of emotions that suggests that our standard emotions are shaped by a narrative arc from beginning to ending with a release. Ngai's interest is in what she calls the ugly (aesthetic) emotions, feelings, and moods that arise when the normal narrative arc of a standard emotion is interrupted. Ugly feelings emerge when agency is suspended and when no cathartic ending is possible. The ugly feelings, which Ngai sees as associated with capitalism, are stagnant ("narrative stasis") and do not allow for emotional release or redemption.[69]

Fear, for example, is a standard emotion in Ngai's sense that responds to an external object and leads to avoiding or taming the object of fear. The related ugly feeling of anxiety, however, lacks a clear external object that one could be afraid about and thus avoid. Therefore, anxiety cannot run its (narrative) course and be conquered. Ngai views anxiety as a basic mood in capitalism. Envy, another ugly feeling, arises as a reaction to inequality but does not lead to a fair equilibrium.[70] Ngai emphasizes the political and collective dimension of these feelings and hopes to unearth their critical potential. Works of art can also aim to generate these feelings as critical ones.[71] In the context of our investigation, what is especially interesting is the idea that ugly feelings such as envy or anxiety reveal the malfunctioning of narrative arcs. Following this line of thought, we can ask what happens when narrative thinking fails. What happens when we cannot close our narrative episodes?

Martha Nussbaum's "political emotions" point in a similar direction. Nussbaum contrasts the positive and productive emotions such as compassion, patriotism, and love (understood as "intense attachment to things outside the control of our will") with negative emotions such as shame, envy, disgust, and fear, which disrupt the positive emotions and interrupt their flow.[72]

A minimal consensus between these studies, it seems to me, is the proposition that narratives that come to an end have a strong tendency toward the positive. Not all episodes end in a clearly positive rewarding emotion. However, even if a negative sequence comes to an end, it is at least over. The work of mourning can begin. Even the interrupted, incomplete, and stagnating narratives produce an index that they could lead to an ending. This thought leads us to the next chapter.

Emotions reward our narrative thinking. Emotions are the mental carrots we chase when we engage in narrative thinking. We hope to get that carrot in the end and eagerly contribute to the narrative in its moves toward that end. Narrative thinking thereby has a transformative power. When the state of affairs is not good, our narrative thinking has us say: the state of affairs is not good *yet*. In February when the weather is nasty, cold, and wet, we can think that spring is coming. We cheer up our best friend after a breakup and tell him he will find someone better. An episode from our lives that is actually quite embarrassing can suddenly be rendered with humor when we tell it to others, and laughter is our liberation. Instead of dwelling on misery, narrative thinking manages to see the dawn on the horizon and has us march toward it.[1]

In the previous chapter, we focused mostly on the individual experiences of people. We considered what the audience of a gossip story or a movie may experience in the process of the narrative. However, narratives can also have an impact beyond the individual level and guide group expectations with grand and collective narratives. This is most evident in the collective narratives of crisis. This chapter will build on previous ideas concerning emotional reward structures and focus on the therapeutic effects of narrative thinking.

Here is the claim that structures the chapter: Narratives exist because they are needed, and they are needed because they themselves contribute to the overcoming of crises and do not merely represent their overcoming. A successful collective narrative is not only a description of the crisis but is itself the vehicle to overcome the crisis. Narratives offer a reprieve, perhaps resolution, by giving form to the disaster.

We will begin with four specific assertions that can be made in regard to collective narratives:

1. The narrative itself performs a part in meeting the crisis—it does not just describe but rather accomplishes something.
2. Emotional arcs are more central than causal arcs (but both can go hand in hand).
3. The narrative makes the offer of a (perhaps new) identity for those concerned.
4. The narrative evokes the possibility that everything could have turned out differently but at the same time cements the given course.

The underlying supposition here is that collective narratives have a therapeutic function that is close to the reward structure of emotions described in the previous chapter. The potential to overcome a crisis also offers a strong incentive for engaging in the story, similar to the hope for a rewarding emotion. The difference is that here a mere easing of the pain of the crisis can function as a sufficient reward.

If this idea is correct, it also shows that the claim that one needs a better narrative often falls flat: not every company wants or needs a narrative, even if marketing departments would often like to have one. Collective narratives require a crisis. (This works well for politicians; their task is to first find or invent a collective crisis and then offer themselves as the solution to that crisis.)

This chapter focuses on the therapeutic aspect of collective narratives as their defining feature. To be sure, scholars have made many helpful suggestions as to what marks a collective narrative. Albrecht Koschorke says that grand narratives and collective narratives develop a "gravitational force" and pull other stories into their orbit. A moral guiding function is also an important aspect, since many narratives have a reward/punishment structure. The collective creation of meaning is frequently recognized as another function of narrative. Collective narratives are often quite vague and flexible, which might make them adaptable to various circumstances.[2]

Narratives play an important role in the process of forming individual and collective identities, as noted by Émile Durkheim in his work on rituals that consolidate communities. In Germany, for example, where the German-speaking people had suffered from an absence of an identity for much of their history, a pivotal moment occurred in 1455 when Poggio Bracciolini published *Germania*, a previously lost text by Tacitus. This narrative gave the German-speaking countries a common identity for perhaps the first time, since they were now allowed to see themselves as the successors of the thoroughly cultured and level-headed people that Tacitus described. They immediately used this notion of identity to assert themselves vis-à-vis other countries.[3] The identity narrative of America, stemming from the European colonizers, was the myth of the New World. This myth not only excluded those people who inhabited the lands before, it also implied a crisis narrative of the Old World, the European homelands that suffered from repressive social hierarchies. Creating the myth of a new and unoccupied land was, even for those who never traveled, a helpful remedy to think beyond the limits of their existing world.

September 11th

How can narratives themselves overcome a crisis? A look at a case study can help here. Immediately after the terrorist attacks in the United States on September 11, 2001, a narrative of the attacks appeared in American media that acquired an astonishing consistency in the months and even years to come—a collective narrative of 9/11. This narrative was adopted by the majority of daily newspapers from the *New York Times* to the tabloid press, showed up in television coverage, became the pattern for most cinematic documentaries that quickly emerged and aired for more than a year, and dominated the picture-rich books that flooded the market and sold in large numbers.

The narrative that emerged and exhibited such "gravitational pull" was modeled on a Freudian trauma therapy. This does not mean that Freud was right in his description of trauma, but rather that his codification of trauma has become accepted as a pattern that journalists and media makers, including those in Hollywood, have in mind when they tell stories. In this narrative, events are presented from the victim's perspective and follow a sequence from shock and repetition to processing and memory. The media in 2001 and 2002 to a large degree cemented the sequence of this narrative to such a degree that the term *ritualization* may not be an exaggeration. This means that the sequence of events becomes prescriptive and starts to function as a collective recipe, at the beginning of which is the victim and at the end the healing and elevation of the collective.[4] Here are the steps of this narrative:

1. Traumatic shock overwhelming the ability to process the events (first-person-perspective images of the Twin Towers in flames, terror on the faces of observers in the street, sudden loud explosions);
2. Victims, death, injury, suffering;
3. Repetition of the shock, initial inability to process shock, involuntary recall, flashbacks;
4. Dust, states of confusion between the present and past of the trauma, afterlife of the traumatic event (ashes, dust clouds, confused faces);
5. Exceptional heroes emerging from the dust who act decisively (firefighters who boldly run into danger, past those who flee);
6. Focus on the healing of the victims and their empowerment (relatives, comforting);
7. Placing a dividing line between past and present, memory presented as a healing strategy, attempt to remember the past event "correctly" and thus discard it (the flag on the rubble of the Twin Towers, candles);
8. Collectives, assemblies, memory ritualization, speeches;
9. Symbolization, religious overtones, totems (again the U.S. flag).

Freudian trauma theory assumes that trauma massively impairs the work of consciousness, insofar as it no longer knows how to distinguish between the past and the present. In this context, Freud describes consciousness by means of the metaphor of a membrane protecting the unconscious, which normally repels experiences and makes them rememberable as past. In the case of trauma this membrane is penetrated, so that external impressions directly enter the unconscious. This has the effect that present and past are no longer separated, and past events remain "traumatically" present in hallucinations and flashbacks. Freud coined the phrase *repetition compulsion* to describe this effect. Healing, then, comes in the form of memory work: the events must be recognized as past and thereby put in place and discarded. The membrane between present and past must become impermeable again. Victims of trauma are healed when they can tell a story of their events with a beginning, middle, and end, because then present and past are sorted again. By means of this storytelling process, survivors place the story outside of themselves.

To reiterate, the issue here is not the correctness of Freud's theory, but the pattern of order it creates. The media representations of 9/11 follow this pattern and heighten it once again by staging it performatively: the media themselves became traumatic insofar as they showed horrific images of the attacks again and again. They emphasized the moment of shock in the confused facial expressions of the people in the photos and thereby also deepened it. The media took over the role of repetition compulsion. On that basis, the media can now advertise and offer themselves as therapists: the (confused) people need the media because they can offer a therapeutic narrative to help. Following the Freudian book, the media present memorialization as the solution to the crisis.

This is a key point: the media themselves perform and become the work of memory that they prescribe to the people. Images of collective remembrance—for example, candles or flags at the sites of the attacks—are offered as a way of processing the losses. The entire sequence from shock to the erection of spontaneous monuments and flag celebrations is ritualized as a fixed sequence of events. According to this sequence, the crisis is overcome when monuments of remembrance are erected. Remarkably, these narrative and media representations are themselves works of memory, since they depict the past. They can thus claim themselves to be part of the therapy and healing of the crisis. The *New York Times*, for example, which garnered a lot of praise for its project to collect the biographies of all victims, was presumably carried along by such a therapeutic impulse. This can rightly be praised and lauded, insofar as the biographies of the victims show readers the enormity of the suffering of the attacks. It is no longer a matter of mere numbers, but of many individual people and survivors whose lives were destroyed or fundamentally uprooted. But, at the same time, one can also be critical, because the newspaper and the media take on a role that is not theirs: the role of therapist. This idea of the media as therapist cultivates an attitude and expectation among media users that puts them at the media's mercy.

Many aspects of this trauma therapy are suspect here, not only the staging of the American flag and patriotism as a cure for trauma. (The flag takes on the role of the Freudian skin or membrane.) The sole emphasis on the victims has led critics to question whether the United States was whitewashing itself with this narrative, insofar as it also acted as a perpetrator in the Middle East and Afghanistan at earlier times, as Noam Chomsky, among others, has pointed out.[5] The media's self-stylization as therapist is also troubling since it simultaneously casts the audience as in need of therapy and thus immature.

Given the one-dimensional response by the media, we should consider what other narratives might have been possible—for instance, collective paranoia or desperation. It would also have been possible to focus on plurality. Even an event like 9/11 has more than one story and perspective. Another variant could have been a form of reporting that deemphasizes emotion and instead focuses on facts—what is known at which time. A more frightening variant—which was certainly also to be found after 9/11 and was propagated by President Bush—is the revenge story, which aims at naming the perpetrator Osama bin Laden and Afghanistan. This narrative leads to the exercise of revenge and destruction. The positive effect of the Freudian trauma narrative was that the revenge story did not become the immediately dominant story. Even George W. Bush waited to build up an international alliance before the invasion of Afghanistan.

Two aspects of narrative can be emphasized on the basis of this example. First, a narrative of a crisis not only orders the events and directs them teleologically toward overcoming the crisis but is itself part of this overcoming. In the case of September 11th, the narrative was offered as a therapy for proper and healthy memory that could overcome the shock. Second, narratives can thereby stabilize the shaken self-confidence of the survivors by offering a new identity or new explanations of the events.

Narrative Therapy

What is the therapeutic dimension of individual and collective crisis narratives? How do narratives contribute to overcoming crises?

Therapeutic aspects have certainly been intimately connected with the development of narratives since early times and are so interwoven with them that it would be nonsensical to see a beginning of therapy in modern practices such as Freud's psychoanalysis or narrative medicine. Narratives are inherently structurally therapeutic because they are explanatory, integrative, resolving, and emotionally rewarding.[6] Myths and religious stories certainly play an important role here, as they exemplify the meaningfulness of actions and experiences to individuals and allow them to participate in them. As narratively thinking people, as *Homo narrans*, we are likely better prepared to process the mental dimensions of crises. It might be possible, to be sure, that we initially experience the crisis as deeper than other, non-narrative beings would, and in this respect we may well suffer more intensely. However, even then we are more likely to transform the crisis event and integrate it into our lives. In this sense, Dan P. McAdams and Kate McLean have described the therapeutic dimensions of narratives as "redemptive" and examined how even traumatic experiences can be transformed by means of narratives into more positive versions of the self.[7]

Narrative therapy places its hope in narration: narration is said to be able to turn around how previous events are processed and remembered and thereby reformulate them in such a way that the crisis becomes part of a coherent, less devastating, future-oriented sequence. This can be done in different ways. There are five strategies in particular that can have a powerful effect.

1. *Narratives can open up a glimpse of a different future.* Even in a state of disaster, a future is conceivable. Many successful Hollywood sagas begin in this way. The *Star Wars* saga, for example, starts in a desert refuge with a few scattered settlements of survivors; the audience gradually realizes that we are in a postwar world controlled by the victors, the evil empire, according to the 1977 episode (produced and released before the prequels were later added). The vision of a differ-

ent future may take a while, but when people suffer a major crisis, loss, or failure, the new beginnings can often be found in the preceding events. The preceding crisis is not completely reversed—this would certainly be an impossibility, since events are indeed irreversible—but an unexpectedly different future may appear.[8] Often, hopes for the collective emerge via the portrayal of an exemplary individual who moves on. New life emerges from the ashes.

2. *The narrative events can be arranged in such a way that the sequence brings about a therapeutic effect.* There is not one particular arrangement of narrative elements that is generally accepted as being therapeutic. However, some of the narrative patterns we described earlier stabilize narratives and thus give them a more harmonious shape. This includes the arrangement of events according to the pattern of beginning, middle, and end; the turn from active to passive or passive to active experience; or the matching beginning and end in the comeuppance tale. In all of these cases a narrative arc emerges that is in itself perceived as productive, cathartic, or redemptive. The narrative pattern itself can overcome the formlessness of the crisis.[9] Benjamin A. Rogers and his team suggest that the arrangement of one's life along the pattern of the hero's journey renders the perception of one's life as more meaningful.[10]

3. *The narrative voice can become so present that it alone gives a strength that indicates the crisis is not the end. The narration itself has the final word, and the strength of the voice of narration triumphs over the crisis.* Many forms of narration draw attention to the voice of the narrator and underscore the impact of the narrative voice. Novels (and movies too) may have implicit or explicit narrators. Everyday stories and social chitchat are also usually told to us by someone, and the way this person relays the events can strongly contribute to the way we process the information conveyed. A mother telling her kids about the ongoing or recent war, for example, will have a profound impact on the kids by the various tones of her voice, which could suggest desperation, rage, or hopeful resolution. Likewise, when a therapist reacts to one's account by retelling it, the confidence, empathy, and probing of the therapist play a major role in how one processes one's own story. Many institutional practices include a strong narrating voice that is

codified in a practice. This is the case in jurisprudence and religious discourses.[11] The word of law and the word of God are narrative voices that firmly present the final word and distribute sanctions. These voices supersede and overpower the negativity of the crisis by their firmness. While these voices are not typically described as narrative voices, they frame and in that sense narrate events that occurred, thereby giving these events their voice of authority. Institutionalizing voices of power, such as those of the law and religion, may have been a way to put to rest painful past events.[12]

The strength of a narrative is also used by the media, whose commentary is superimposed on the facts presented. It is not that the institutions or media simply add a concluding verdict, but rather that they first give the entire account a form by adding ongoing commentary. This becomes clear in the case of Elizabeth Holmes, the founder of the failed company Theranos. Holmes had founded her company with the help of gigantic sums from investors and claimed that the devices she had developed could provide highly cost-effective medical diagnoses. However, it turned out that this claim was a lie, and the company produced its blood tests by means of conventional, more expensive devices from other manufacturers. Holmes, who became a billionaire as a result of the hoax, was subsequently in the dock. The case attracted a great deal of public attention. Many things came together: unscrupulous swindling at the highest level, the culture of Silicon Valley, a fascinating protagonist casting herself as a female Steve Jobs, as well as possible sexism by the public and media, which dramatized her conduct more than similar incidents of male hubris.

Exemplary cases of this kind often acquire a collective dimension. In this instance, the temptation of investor capitalism is narrated and negotiated. The crisis is capitalism, and now a second drama is added, in which the previously active agent of this capitalism becomes the passive recipient of a verdict. What is always negotiated along with stories of this kind is *whether* there is a narrative agency here that knows how to negotiate this capitalism, and *whether* there can be a hope for narrative therapy. It is the drama of whether a narrative position is possible and has power over events. Depending on the opinion of the commentators, the entire story of Holmes's rise and fall can be described as the case of an unscrupulous imposter, in which even her pregnancy is an attempt at trickery, or the case of a hypocritical witch hunt in which the female founder is held to different standards than her male colleagues, who get away with their lies. In both cases, the narrative voice of the commentator carries the narrative and assigns it a position of strength whose judgment is above the crisis.

However, strong narrative voices do not rely on existing established institutions, such as law and religion, or institutionalized practices, such as commentary by the media. Rather, strong narrative voices *institutionalize* themselves in the minds of the audience. Institutionalization here means that this voice can be mentally reproduced, sustained, and activated to provide stability in the way stories are told. Simply put, people can mentally call up this voice and have it frame and explain events. I can use the voice of Obama to frame events for me, I can retell a difficult situation by imitating the way my friend would recount it, or I can experience an event in the way the narrator of *The Catcher in the Rye* tells his life.

This was the insight provided by Gérard Genette's narratology that every story is told from a perspective and with an implied narrator.[13] Anyone who reads, say, Leo Tolstoy, Gustave Flaubert, Virginia Woolf, or Thomas Mann gets carried through a narrative by the voice of its narrator and can examine the events from the stability of that voice. We can morally scan events as Tolstoy does, cynically observe like Flaubert, engage in interior monologues in the style of Woolf, or take an analytical position as in Mann. I am counting on my readers to insert their own experiences at this point, which may go beyond my Westernized canon of reading experiences. The strong voices of contemporary literature have comparable power, each clearly borne of what these narrative voices include or exclude. Michel Houellebecq, for example, leaves no space for his characters to acknowledge the humanity of any other character. Sally Rooney breaks with the idea of consistency and shows that the voices of the respective characters can suddenly give way, leading them to do exactly what they had previously ruled out. Such literature invites its readers to continue spinning the threads of the narrative by extending the narrating and framing voice.

When crisis erupts, the strong personalities in our lives take a leading role in shaping our response. In moments of important decisions, we may find ourselves asking how our grandmother, our pastor, or our mentor would characterize the situation. This is not about what these guiding figures would actually do, but rather how they would tell and frame the events.

The paradigm of a strong role model has a long-standing prominence in the history of moral philosophy, but it has been neglected in recent centuries and rarely appears in current moral psychology. However, this is precisely what happens in many collective narratives: the voices of individual public figures from politicians and intellectuals to celebrities and influencers can continue to run in our minds as narrative role models. They are influential not only in what they say specifically but in how we continue their speeches and use them to tell ourselves out of crises. Donald Trump, for example, continues to speak from many minds in the narrative he has fabricated.

4. *In the case of traumatic crises, the mental repetition of the crisis and representation in narrative form give hope that one will be better prepared in the future for similar events and will be able to overcome past trauma.*

The key here lies in the benefit of repetition. Narratives, by offering simulations in an offline mode, allow us to process past events and prepare for future events. In many theories of trauma, both dimensions of this process—better preparedness for the future and overcoming past trauma—are connected. In *Beyond the Pleasure Principle*, Sigmund Freud proposed that repetition of past trauma allows us to project it into the future, thereby allowing us to try out newly acquired defenses.[14]

Let me add a historical note here. Interestingly, the three benefits of narratives discussed above have been known and utilized for a long time. This fourth one is, in historical terms, a much more recent discovery or perhaps invention about the psychological apparatus and its connection to narrative structures. In fact, the theory of psychological trauma (and also of post-traumatic stress disorder) was first developed in narrative literature and not in the medical sciences. The physicians of the Napoleonic wars around 1800, who must have faced a large number of deeply traumatized veterans and survivors, simply diagnosed their patients with a lack of masculinity. These physicians suggested that the war merely brought out the true and despicably weak nature of these unmanly soldiers.[15] At the same time, however, a number of literary authors of the classical and Romantic periods, beginning with Karl Philipp Moritz in Germany, developed in their narratives the idea that overwhelming impressions can shape an individual in such a way that the person will subsequently be haunted by perceiving returning patterns of powerful impressions over and over again, thereby involuntarily reliving them.[16] Literary fiction with its awareness of returning patterns discovered a structure here that anticipated many elements of post-traumatic stress disorder.

The intellectual co-history of narratives and trauma has been in many ways deeply intertwined ever since. It should be noted that literary authors could make a career out of the idea of trauma. Trauma allows the arrangement of narratives in such a way that later actions remain related to earlier ones, and the narratives thus gain coherence. Many a plot since 1800 explains the development of a character by linking his or her future actions to past trauma. This unfolding of trauma brings narrative coherence to biography, since one's past offers a blueprint to understand and explain one's present state.

Already in Honoré de Balzac's story "Adieu," the trauma narrative is given a therapeutic dimension. There, the attempt at trauma therapy occurs at the level of the characters. A young bride is separated from her fiancé during a battle in the Napoleonic war and subsequently loses her mind. Her fiancé survives the war, but she does not recognize him later. In order to heal her, he decides to reenact the violent battle for her (with the help of many dressed-up farmers) to restage the couple's moment of separation and provide a better ending in the eyes of the bride. Indeed, she comes back to her senses and recognizes him—but is killed by the intensity of the experience.[17]

In the literary idea and formula of the trauma narrative, the terrible first event becomes a script for later repetition. What was then a temporal order of events becomes prescribed with causality. This pattern is present today in an endless number of detective stories where criminal behavior is explained to the audience by some traumatic repetition performed by the criminal. Hollywood often reveals "the truth" of a character in the repetition of a childhood scene or trau-matic experience. As the story progresses, readers and viewers learn what actually took place at the beginning; this revelation explains the puzzling actions and thus gives the narration coherence.

Needless to say, these lines are not intended to trivialize the suffer-ing of traumatized survivors as mere fiction. The point here is the large pattern of achieving a potentially therapeutic effect by means of narratives. The trauma narrative can become therapeutic for the au-dience (as opposed to the trapped characters and survivors) because they can embed the horrific event in a series of repetitions. What is already therapeutic here is that the narrative continues. The devas-tating catastrophe is, despite everything, not an end point. It can be told and thus already confirms a narrator's position—from the per-spective achieved after the crisis. The narrators here do not have to pull a rabbit out of the top hat's black hole. Rather, the mere presence of the narrator is testimony to the fact that the narration does not get out of shape—and that would be the real catastrophe, if the narrative could no longer be told.

5. *Narratives allow us to retell events differently.* This fifth strategy is the common structure of the previous four. They all allow a retelling in different ways by focusing on or adding a strong voice, providing closure, creating new sequences, opening up hopes for a different future, and building up resolve. The (therapeutic) narrative offers a second rendering that follows the first narrative of the crisis event. I place the word *therapeutic* in parentheses here because in a way every retelling and every narrative has a therapeutic effect, as I will suggest. Verbalizing an event and thereby retelling it or retelling an already existing narrative gives form to the first narrative of the crisis event. Goethe's words, already quoted, that it is "pleasant" to "remember some obstacles," also apply here.[18]

To be sure, the retelling or second narrative does not have to lead to a happy ending. There is no happy ending to the global financial crisis of 2008. But there is a collective narrative that very quickly took hold at the time and identified the culprits as the unscrupulous and greedy U.S. bankers who had issued dubious loans and mortgages. In this case, the very naming of the culprits by the narrators has a therapeutic effect, because naming and thus exposing and blaming them is already a punishment and provides the rewarding emotion of satisfaction.

These narratives involve a shift from active to passive or passive to active experience, as discussed earlier. In the September 11th attacks, "America" is first the passive and traumatized victim but then works its way back into an active position. Similarly, in the global financial crisis of 2008, the world is victimized but finds active retribution in naming the guilty. The villains, on the other hand—that is, the bankers—become the passive recipients of shame in the narrative. Of course, further legal prosecution is also expected and becomes part of the retelling, as in the case of Elizabeth Holmes.

Missing Narratives: The Covid-19 Pandemic

I began writing this book on the day when the Covid-19 viruses had definitely arrived on the North American continent and the tentative attempts to contain the outbreak were declared a failure. I had just given a talk at Stanford University using the still-fresh empirical data from Chapter 3 and was on my way back to Indiana when it was announced at the airport that the virus had arrived in Palo Alto. On the plane, my writing began, and behind me the airport was closed. I don't mean to equate the growing lines of my book with the increasing spread of the virus, but this coincidence explains in part my interest in how a narrative can emerge from an international crisis.

After my return to the Experimental Humanities Laboratory, my colleagues and I decided to collect stories about the pandemic. We assumed we would catch the development and spread of a collective narrative. One of our approaches was to ask participants in our experiments to tell true and fictional stories about the virus. Here is a relatively typical story from March 2020, the time when the pandemic began.

> I recently went on a cruise about the time the coronavirus was starting to spread from China. We were hearing reports about other cruise ships being stranded and having sick people on board. I was terrified that we would become another stranded ship.

I thought about being quarantined and having to be away from family. I was deeply worried that I would get the virus and end up in the hospital, or worse. I thought about how much my life had meant to me up until this point. I thought about what I wanted to accomplish in life once we were safely back home.

Fortunately for me and my spouse, our ship was not one of the ships that needed to be held in port. We were allowed to disembark on time. Even though we were lucky to have avoided the virus on the ship I am still concerned, and paranoid, that I will contract the virus from others.

Here is a short narration from March 2021, a year later:

Man, the coronavirus pandemic was crazy. I honestly felt I wasted a year of my 20s just because I couldn't go out and hang out with my friends from college. It was pretty bad here in Texas and it didn't help people didn't want to wear masks. It took them such a long time to get vaccines out and even when they were out, only a limited amount of people could get vaccinated initially. I struggled a lot with mental health because staying at home does that to you. I'm just glad it's finally over and I can hang out with friends and family again.

The comparison between the two narratives shows clear similarities. In both, we get a sense of being trapped, one imagined on the cruise ship and one taking place in Texas. But the description of the crisis is different. The self-encounter on the cruise becomes a symbolic hibernation story in which the pandemic period must be survived like an icy hibernation. On the cruise ship, there is a potentially meaningful and existential encounter with one's own life and death: "I thought about how much my life had meant to me up until this point. I thought about what I wanted to accomplish in life once we were safely back home." In the March 2021 narrative, on the other hand, the mental crisis is not given a spiritual dimension but is simply labeled and normalized: "I struggled a lot with mental health because staying at home does that to you." This is a tale of a lost year. The 2020 narrative has an emotional arc from fear to gratitude and ongoing paranoia. The narrative from 2021 is unemotional; the crisis is simply written up as negative and its end as positive. Powerlessness dominates both stories, but in 2020 there is a residue of potential for action as the narrator faces her life and thinks of her goals. In 2021, only grammatically negative expressions are used for the pandemic, and even the protests are classified negatively only in a half-sentence, "it didn't help people didn't want to wear masks." There are no heroes, only a vague "they": "It took them such a long time to get vaccines out," which probably refers to the government and the health care system at the same time.

Of course, these are only individual narratives, selected by me, and not collective narratives. But they indicate a major line of development: in 2021, the personally perceivable story with good or bad luck and with potential for meaning becomes a simply negative story about a lost year with no gain in experience. This narrative is therapeutic only in the sense that the narrator has chosen a perspective of hindsight indicating that the episode has come to an end.

Here are some more of the narrative patterns we observed:

- a symbolic hibernation full of privations with a return to normality ("it was bad, now it's over, we want to forget it quickly");
- an individual story of suffering, how someone got the virus and either survived or died;
- an attribution of blame: China created the virus or deliberately delayed issuing warnings;
- various other conspiracy theories—for example, that the government shoots tracking devices into all people along with the vaccine;
- an attribution of failure: those in power reacted irresponsibly or poorly;
- a denigration: if it were not for the opponents of vaccination, the virus would already be defeated;
- a division of people into binary groups, such as the reckless and the cautious, the happy and the desperate, the carefree vacationers and the fearfully bunkered;
- the failure of divine justice: someone got the virus, even though he was so careful;
- a hero narrative—for example, one that celebrates the developers of the novel vaccines;

- finding happiness in one's own home, alone or with family;
- celebrating that it is now over and we can all come together again;
- anger at the imposition of those in power that one must continue to wear masks or face restrictions if one remains unvaccinated.

Moreover, certain motifs kept recurring. In the beginning, people noted the emptied ghost towns, the lack of toilet paper after people bought it up, and the miseries of homeschooling kids. Heroes, in contrast, were largely absent. Later, new motifs were added, such as the man wearing the mask under his nose, jealousy toward people who already were vaccinated, the first celebration with friends and family "after" the crisis, or lying about things like vaccination status or symptoms.

However, most of these elements and stories do not qualify as collective narratives. The most common account in our data remained the symbolic hibernation pattern. It barely counts as a narrative since it meets hardly any of the narrative conditions we have so far identified. It often lacks a switch from active to passive or passive to active; it does not segment an episode; it names no protagonists; the events are weak; and there is no real emotional arc here, especially in the sense of a rewarding emotion. And this means the narrative most likely does not contribute to the resolution of the crisis.

In contrast to the hibernation story, the villain story ("China"), the conspiracy stories (those in power . . .), or the hero stories (science saved us) would have offered more of a typical collective narrative that would not only have given the crisis a clear spin but also contributed to overcoming the crisis by moralistic finger-pointing. The 2008 global financial crisis with its clearly identified culprits was a different story. Another possibility would have been a literary immunization paradigm that prepares and immunizes people to some degree by having them imagine and thus co-experience a threat before they encounter it, as Johannes Türk describes this paradigm.[19] However, such an active narrative engagement also did not materialize.

The hope that my collaborators and I had to observe the nascent state of a collective narrative has not been fulfilled. Even several years later, long after the official end of the pandemic, there is no strong collective narrative circulating. It should be acknowledged, of course, that there are stories galore, but the bar for a collective narrative is set high here. This absence of a collective narrative distinguishes the pandemic from many other crises—such as September 11, the financial crisis of 2008, and the fall of the Berlin Wall—where narratives quickly took hold.

A historical case also belongs here. Historians often cite the case of the highly destructive Lisbon earthquake of 1755 to point to the emergence of a novel and therapeutic collective narrative. However, Gerhard Lauer argues, based on archival work, that the earthquake, despite its devastating effect, did not establish a new narrative. This is contrary to many former assumptions. Instead, the earthquake was simply incorporated into existing theological explanations.[20] In short, we do not find collective narratives for every crisis. Sometimes, suffering remains unspoken.

What Have We Learned?

In this chapter, we considered the therapeutic effects of narratives with a focus on collective crisis narratives. Narratives mitigate or even settle crises by providing retellings that offer hope, open up new perspectives, point to the future, or give the stability of a frame. The key element to emphasize here is that narrative as such becomes part of the solution or therapy. The narrative repeats the crisis event and in doing so offers tools to examine and retell it differently. Thus, the crisis can become a turning point. Moreover, in this continuation of the crisis as a narrative, an emotional reward beckons that unfolds therapeutic aspects by helping to classify the crisis, to limit it, and to gain from it a potentially positive dimension. However, there is no simple recipe here, and in many collective crises such a narrative fails to materialize at all.

Some readers may raise the objection that there is also the opposite case of harmful narratives. There are indeed times when a narrative instills misery in us. When a jealous spouse imagines her partner cheating on her, repeating this narrative mentally does not immediately offer resolution. Instead, what began as a quick thought or suspicion can become a mental reality by means of the retelling. Likewise, someone in a state of depression may be stuck retelling himself the story of his failures. An example of a harmful collective narrative would be propaganda that galvanizes negative sentiments and utilizes them for enraging its audience. We will have the opportunity to return to these forms of negative narrative sentiments and stagnations in what follows.

What are others to us?

What are other beings in our mind?

How do others manifest in our thinking?

What kind of knowledge do we have of others?

What kinds of ideas do we gain about those who appear in narratives?

What are the cognitive operations by means of which we create and approach characters in narratives?

In what follows, we will explore the question of how we create characters in our mind. That means we will ask which mental operations we use to generate and approach characters. I will propose that there are three mental operations that have particular importance in this respect—namely, *tracking* and *playing (to be) them* and *justifying* their actions. For an initial orientation, these three operations can be outlined as follows.

Tracking: The character did *X* yesterday and I remember it.

Playability: I can place myself in the character's shoes and do actions for him or her.

Justifying: I can explain and thereby to some degree justify why he or she did something—"The character did *X* because of *Y.*"

These mental operations, I propose, can explain how we create narrative figures and characters in our minds with relatively little cognitive effort.

These three operations do not correspond to the conventional categories of narratology, since I do not start from existing works of fiction with their ready-made characters and classify them. Instead, my start is again from the mental operations by which we engage our narrative brain. I also believe that these operations matter for both narration in fiction and the ways we think of others in real life.

Still, these proposed mental operations have some proximity to narratological considerations. Matías Martínez, for instance, in order to mark the difference of fictional characters in narratives, has suggested that they are "more complete and at the same time more incomplete than real persons" because one can find information from them only within their narrative world.[1] Although Martínez applied this insight to explain the difference between real people and fictional characters, one could also turn it around and suggest that it describes what happens when real people enter our narrative thinking: once real people are the subject of narratives in our mind and in that regard become characters, we track them, play them, and justify their actions.

Tracking: On the Genesis of the Construct of Person

Moby Dick left a lasting impression on Captain Ahab not simply because of his magnificent appearance but because he had robbed him of a leg. This impression motivated the one-legged hunter to chase the whale across the world's oceans. Because the mysterious whale stood out by color and size, Ahab could not only identify but also individualize him, making him the object of his vengeance. It is precisely this combination of identification and emotional charge with which we approach other beings.

First of all, we're very good at identifying individuals. When we have a group of people in front of us, we usually know very well which of them we've met before. In general, it's a great pleasure for us to look at faces, as the many portraits in art prove. In fact, there's probably nothing we observe and remember with such intensity as faces. There are, of course, good reasons for this. It is particularly interesting that we value objectively small differences between faces very highly. One can quickly see this with a simple experiment: if we make small drawings of a set of faces, they all end up seeming very different to us. Some we like, others we don't. They may differ by only a few strokes, such as dissimilar eyebrows or corners of the mouth, but nevertheless they have a different effect on us. We can identify people by their faces, as well as their voices, smells, and typical movements. And that means, first of all, that we can recognize them. If a person has done something bad to us once, we don't forget it, and even after years we are very careful when we see that person again. Conversely, I also remember very clearly when someone once seemed interesting to me, but I didn't get the opportunity to talk to him. I then hope for later reencounters.

Evolutionary biologists speak in this context of our ability to *track* others, that is, the ability to mentally mark individuals over time. We can quickly put together the times we have met someone over the years and build a composite of her overall personality based on these few encounters. We still know now what someone did to us years ago. We also remember our judgments of people well and can quickly activate them again.[2] Some people become warning signals to us in this way; we react quickly and have the option to flee.

Tracking is a fundamental adaptation without which our social communities could not function. Social obligations, morality, responsibility, and legal accountability could not be imagined without some tracking. At a minimum, tracking implies that we can recognize a person or character and relate her past and future actions to each other. For us as observers, she is the same person in the past, now, and in the future. Because of this, we can blame or praise her. Because we can track others in our mind, they can have the appearance of a strong identity for us. A character we track in this way has a constant contour for us and we can hold her accountable. This then also means that we hold the person accountable in the future for what she does now or did a long time ago. This accountability thus aims at a core of stability and self-similarity over time.

Tracking and identifiability play an important role in the formation of cultural institutions. Identifiable individuals can be granted privileges and given status.[3] Legally, they can be prosecuted and be held liable. This assumes that something essential about them remains stable over time. When individuals are rewarded or punished, two narratives are created, which are then joined together by means of identifiability: the narrative of their relevant action and the narrative of the (actual or expected) sanctioning (reward or punishment). They are held together by the tracked individual. What this means here, then, is that the temporal difference between the two narratives disappears in the sameness of the character. We usually take this for granted, but we should still pause at this assumption of a lasting identity. We generally do not assume that people have more than one mind and we insist on the idea that the different states of a person and different identities are still encased in one overall trackable character. For humans, past acts are not forgotten but remain attached to the person later. Past acts have an aftereffect—except in cases that are institutionally classified as expired, and we have complex ways and rituals to declare past acts as waived, settled, or forgiven. We also generally make exceptions for children. Tracking and the connected idea of mostly stable identities are a central element of human culture and our institutions, since these always aim at regulation and perpetuation of similar behavior.[4]

Josef Perner, a psychologist working on theory of mind research, and his colleagues have developed a model according to which we cumulatively, and thus step by step, acquire more information about other people in order to better predict their inner states and actions. According to this model, the "mental file theory," we collect knowledge about objects (or "referents") each in a kind of mental folder and then assign this knowledge to specific people (A thinks X about object C). Correspondingly, we can also consider other persons as "file folders," which we fill with information and thereby track them.[5]

In this way, we may develop an understanding of narrative characters and real people because we bundle information about them—that is, we track them over time. There may well be surprises in the process. It may be that someone has different characteristics than we have been able to observe so far. We might mentally collect information about objects in two different folders until we realize that they are the same object or character (Clark Kent and Superman are one person).[6] People can change, or the writers of a television series can decide to give a character new features. But in most cases, we are probably satisfied with the ideas we have about another's identity and revise them only when it becomes necessary. Tracking allows us to develop a range of expected behaviors from a character. The reward of tracking is the character's predictability.

Errors happen. Tracking takes place from the outside perspective of someone who follows a character. This observer assigns past deeds to this character accordingly, which in turn can lead to predictions. A person who did something generous yesterday, for example, may then be classified as a good person, although such a judgment may prove incorrect. Consider, for example, the deliberate deceptions featured in Samuel Richardson's *Clarissa* (1748). In this novel, the rich libertine Lovelace continually tries to seduce and assault Clarissa. At one point, he hires actors to impersonate his family to present him as a morally upright person to gain her trust, only to seduce her according to his plan.

In Praise of Playability and the Case of Tulpamancy

In 1976, a mostly new genre of literature emerged in the United States under the name "Choose Your Own Adventure." It became a standard read for American youth in the following decades. In these books, readers were addressed directly as "you" and found themselves as characters in the stories. There, they had to make decisions for this You-character about what You should do in the given situation, or what they themselves would do. Depending on the decision the reader chose, a reference followed giving the page number where the story would continue. Those who decided to secretly follow the mysterious visitor, for example, could continue reading accordingly. Or those who wished to follow a previously interrupted plot could continue on that storyline. Today this genre is known as interactive fiction, which has moved on to digital platforms in the form of video games. Digital presentation simplifies the flow without the necessary back-and-forth scrolling as it directly continues the story once readers or listeners make a choice. Depending on the title and the company behind the title, there are insertions of voices, sounds, and movie episodes. Typically, decisions are made not only for just one character, but for several.[7] Several additional features also affect the decisions the readers or players of the stories make. In the story *Detroit: Become Human*, labeled as a video game, for example, one can also look up what other players have decided (by percentage of choices). *Detroit: Become Human* sold more than nine million copies by December 2023, and many players/readers report playing the game more than once to explore other possible stories, as Victoria LaGrange has examined in her research.[8]

In such books and games, the community of players distinguishes between playable and nonplayable characters. The playable characters are those in the story for whom one has to (or can) make decisions. I think this term is extremely fortuitous and would like to employ it here as a fundamental cognitive operation in narrative thinking. Characters become "playable" when we can mentally envision how and what they might do, say, or feel in certain situations. In the same way, real fellow human beings and other creatures can become "playable" to us when we internalize them. This does not yet mean that we have some full-blown empathy for them or create a mental model for them, as in theory of mind. However, we can imagine placing ourselves in their situation, and we consider the array of feelings and actions one could have in that situation. We can relate to these characters from a grammatical first-person perspective.

This kind of "playability" is a great achievement of our thinking. It allows us to recognize another's space of action—that is, we can not only simulate another's actions but also weigh these different actions against another in our minds as if they were ours. We take an inside perspective of the different possibilities of what this character could do in this situation. We do not need to have a clear and articulable idea of the identity of this person exactly in our mind but can simply recognize that he is an independent agent that can act accordingly. To stress this again, this is not yet empathy, theory of mind, or a clear understanding. However, it is a matrix of the other person as an animate being who has the capacity for agency.

This playability has been discovered and enjoyed in recent decades in computer games and the new worlds of virtual reality. Playing characters in virtual worlds now occupies a significant space in the lives of many people; the popularity of games like these suggests that this play in itself is easy for us and often associated with pleasure.

A playable character in our thoughts has a large space of probable and improbable possibilities, constrained only by the specific situation. On one hand, anything goes; on the other, there are consequences for actions (the consequences that we touched on in the previous section on tracking). Playability here, then, vacillates back and forth between an "anything goes" and a "there will be consequences." In yet another way, the character is caught up in an in-between space. We recognize that a character acts independently, but we appropriate this room for maneuvering and mentally make or test possible moves. In this regard, others are thus controllable and uncontrollable at the same time.

I consider the notion of playability to be a convincing alternative for too static or textual notions on characters in narratology and for the too mentally elaborate calculations of theory of mind. Playability is simpler and more fundamental. It should at least be suggested, however, that I also think this playability could be considered as a concept in developmental psychology. For example, even before children have a clear theory of mind—that is, a concrete idea of how specific other people think and feel—they may already be able to grasp the playability of characters. This concept could also be important for an understanding of people on the autism spectrum. As has often been pointed out, people on the autism spectrum can have a clear propensity for narratives, even as they show below-average performance in a range of theory-of-mind tests.[9]

It would also be interesting to investigate whether non-human animals have the capacity for playability. It is still controversial whether, for example, dogs, chimpanzees, scrub jays, octopuses, or dolphins have the capacity for theory of mind. Perhaps it would be more useful to ask to what degree they are able to mentally play other beings and thus recognize that other beings have spaces of agency they could simulate. This would be expressed, for example, by attempts to purposefully influence the actions of other beings. (I can just imagine my cats standing in the house in front of a locked door and whining. In doing so, I figure, they are hoping I will let them outside, which I won't do during bird breeding season, but they want to influence me accordingly. I imagine they are wishing they could play me now.) Perhaps this capacity for playability is within reach for some animals.[10] At this point, I see the need for further exploration of an underappreciated capacity.

For some years now, a new phenomenon has been appearing under the name "tulpamancy." Tulpamancers are people who describe themselves as hosting imaginary beings in their heads whom they have brought into existence on their own and with whom they are in communication.[11] Tulpamancers thus view these imaginary beings as a creation in their heads, which distinguishes them from people with schizophrenia or people who report religious experiences.[12] The term *tulpamancer* emerged in the 2000s in an internet group that was inspired by a practice in Tibetan Buddhism. A *tulpa* in Tibetan Buddhism is an imaginary being or helpful spirit that a monk in danger could summon to assist him. The members of the original internet group willfully tried to imitate this practice by summoning characters from the television series *My Little Pony: Friendship Is Magic*, and soon the phenomenon caught on. Others imitated it or recognized a practice they were already following. For tulpamancers, these tulpas have their own space and their own rights in their mind. Many report that they regularly negotiate with their tulpa or tulpas. Once a tulpa has been summoned, it remains and resides in one's mind.

For our investigation, this phenomenon of tulpamancy is interesting for two reasons. First, it shows us a case where playability precedes the actual character. Apparently, this internet group wanted to spawn imaginary friends, and they apparently succeeded. The will to play and the idea of a space of action for these characters preceded the concrete creation of the characters.

Second, tulpamancy reveals the relationship we can build with (imaginary) figures and characters in our minds. Tulpamancers can imagine the characters at eye level, can communicate, negotiate, and act with them. In fact, many report that at times one of their tulpas acts as a "fronter" who speaks to the outside world and makes the decisions. The act of allowing a tulpa to take on the role of fronter is described as "switching."[13] Similarly, tulpamancers also report how they seem to talk to their characters. In these cases, of course, the tulpamancers do not play their characters but instead allow them to play. Still, it seems plausible to suggest that the playability of the tulpa characters was a point of access to invite them in.

Sigmund Freud once commented that we can not only continue to play certain characters in our mind but, in the process, multiply them so that our soul landscape becomes a populated stage ("Dostoevsky and Parricide," 1928). Perhaps Freud was a tulpamancer too. This makes me wonder what drove him to summon a punishing tulpa that he called the Super-Ego in himself and a mysterious, dark tulpa he called the Id. If he had opted for other tulpas, perhaps by taking clues from watching *My Little Pony*, the history of psychoanalysis might have taken different and maybe more happiness-focused forms.

Indeed, I think this ability to summon characters within us and give them space is not a special case, but quite attainable by most of us—although probably few of us would call ourselves "tulpamancers" and set out to give imaginary characters a seat at our decision-making table. We do not just have the ability to play characters in our minds like puppet masters; we can join the puppets in thought and act with them.

Some people certainly cultivate the ability to create beings within themselves more clearly, perhaps taking their cue from how they actively simulate other people. I personally had a sudden revelation when I thought of my mother, who had recently passed away. Her husband, my father, had died at an early age, which until recently I thought was an accident. (I will take this up in the next chapter, but I now know that my father was killed.) She had never remarried and, after my sister and I moved out, lived relatively happily on her own, doing volunteer work now and then and seeing friends regularly but not particularly often. Now it suddenly became clear to me what the source of her pleasure was—namely, that she continued to live with her husband in her mind and most likely had long conversations with him every day. From her childhood on, she had been an avid reader of fiction and biography, and I remember how she always encouraged me, too, to vividly picture the characters in novels. This, I now think, was also the secret of how she was able to continue living after the murder of our father. She became a tulpamancer. And like the Buddhist monks, she seems to have averted the evil spirits with the help of a tulpa.

I am not in a position to consider here whether the phenomenon of tulpamancy is overrated by either the tulpamancers themselves or their academic observers, or whether tulpamancy should be described as pathological or helpful and curative. Hence, instead of talking about tulpamancy in the narrow sense, it seems important to me to use it to outline a spectrum of behavior that is at least loosely connected with it. In addition to religious practices of visualization (including in Christianity) and the aforementioned techniques of authors to engage in dialogue with their characters, this also includes more everyday forms of turning people and beings who are important to us into characters in our imagination. These can give advice to us in situations of moral ambiguity or in cases of radical uncertainty. Some people simply ask themselves what a trusted friend or positive authority figure would do or say. The imagined figure may respond. Others perceive an inner voice that comes to them as if from the outside.[14] All of these forms include cognitively helpful strategies to go beyond oneself by conjuring mental beings that are a product of mental playability.

There are other figures and characters that are also clearly playable for us, including erotic ideas that we condense into figures and that excite us. What is interesting here is that on one hand we direct and play these figures but on the other hand we perceive them as self-directed and independent from us. Playability of these figures allows us to place ourselves in the middle of our creations. We can become one of them, and we can also be among and between them. We, too, are playable, or at least part of the playable characters. We are able to spend a part of our lives in this kind of fiction and imagination.

The drama scholar Amy Cook has used the term *casting* in proximity to what I call playability here. In a narrow sense, casting refers to the selection of an actor for a role. As Cook elaborates, there is complex cognitive activity involved in casting. An example of this occurs when one imagines, to use Cook's illustration, a black actor performing the role of Hamlet: some may be unsettled by this choice given their prejudices and expectations.[15] Similarly, playability shows that we and everyone else could potentially act in any role, but that doesn't mean it makes no difference who's playing whom. Before we may be able to explore differences between specific people, the fact that we can mentally slip into a role and play others opens up a productive space of exploration of new situations and the spaces that others inhabit. Spaces of such possibility are what narrative thinking is all about.

(To be sure, it would be highly problematic if we would think, based on some playability, that we now "understand" others. However, this shortcut is not the point of the mental operation of playability. Playability is about our mental mobility, a topic that we will discuss in the final chapter.)

Explanation and Justification as the Basis of the Narrative Person

Anyone who uses a dating app to find a partner today is asked for specific preferences, depending on the platform. This makes it easier for other people, but also for the algorithms, to find matches. Presumably, similarities between people—for example, political orientation—are an advantage here. Apparently, many people today expect to find a mirror image of themselves in their partner. But for narrative persons, preferences such as favorite food and favorite activities are only of secondary importance. We collect this information in the course of tracking, but it only provides superficial information. In contrast, what really matters is how we can use this and other information to make sense of others, to explain their behavior and, in controversial cases, to justify it.

In addition to tracking and playing characters, there is a third activity that has a central role in how we deal with characters in our narrative thinking: we can explain and justify them. This justification, as I will argue below, takes an interior perspective with respect to other characters and gives their behavior a legitimacy that elevates them to characters in the true sense of the word.

Justify means that we can give reasons why characters acted in a certain way in a controversial or complex situation. Of course, instead of "justify," we could also just say "explain" or "understand" at this point. We understand, for instance, why a character does something. The accent on justification here is meant to indicate that this kind of explanation occurs primarily when the character is under pressure to legitimize his or her behavior. After all, we do not have to provide explanations for all action continually, but only under certain circumstances where other options were available that some might have preferred. I usually don't have to justify or explain why I'm getting into my car, but if I abruptly end a conversation by doing so and leave my interlocutor stunned, some justification might be warranted. When I witness the actions of others, standard explanations of their behavior are usually available to me, but I only actually have to access the mode of explanation when the actions become strange and problematic. Couldn't she have done something kinder? Why did he say that? That is, the explanation usually only becomes relevant when there is a question of legitimacy.

The pressure to legitimize oneself exists in many situations. In addition to direct legal and moral calls to justify oneself, these include a variety of expectations that we place on ourselves as well as the demands others place on us. Theological and religious ideas also play an important role here. In many cases, expectations emanate from institutions, including internalized institutions such as role expectations: the role of mother, for example, encompasses many obligations and strictures, while the standard set for fathers is usually much less demanding. Our Western culture exerts pressure to prove oneself as an individual ("the self"), to enjoy oneself, and to be financially successful. We often justify ourselves not with verbal statements but through behavior or career choices. It is perhaps not inconceivable to argue that many young people in the Romantic period wanted to become artists because it was a response to the pressure to prove themselves as free individuals, different from others and cultivating a style of their own. High income or the accumulation of wealth can similarly be seen as a strategy to legitimize oneself, because those who have a lot of money rarely have to justify themselves. The need to legitimize, justify, and, consequently, to excuse certainly plays a central role in the constitution of the self.[16] In the following, we will focus on the concrete narrative forms of finding explanations and justifications.

When people do something, it usually has a clear intention that we associate with the action. This is usually unproblematic and we see the action and intention as connected, so we rarely need to consciously think of this intention. Behind this connection of action and intention there are, of course, quite complex learning processes, but they usually have only a background function for narrative thinking—except in cases where justification is needed. In narrative thinking, the important cases are those in which the actions (including expressions and emotions) of a character are not obvious but are instead problematic. When an action is unclear or reveals an intention that we find dubious or wrong, the work of explaining and decoding begins. This becomes more intense the more we care about that person or character and, as I will argue, the more we take the inside position of that character.

When we aim to explain and justify another's actions, we take an inside position of their character. In taking an inside position, the real person becomes a narrative character of thought. In this, taking an inside position builds on the playability we spoke of above. But what is added now is that from the inside position we give a reason why a certain behavior might be meaningful or make sense. That is, we begin to rationalize, legitimize, and justify the behavior *from the perspective* of that character, thereby connecting the act of justification with something close to a theory of mind of the other.

Scholars often speak of a narrative identity and self-sameness. Often a central function of narrative is seen in the fact that we can present a more or less coherent story of how we became who we are and thus also justify why we have a certain identity or behave in a certain way.[17] This is not quite the same as what I mean here by the emphasis on the mental act of justifying actions. The emphasis here is not on unity and coherence, but on making sense of concrete events in our lives at the moment and justifying them as one-time occurrences. Justifying means to explicate why a specific action is congruent with a norm or institution—or in this case with an idea of the self. To a strong feminist, it may make perfect sense to spray-paint a monument against abortion in a country with a repressive regime. At the same time, the very need for justifying expresses and marks a gap: apparently that action remains problematic and in need of elaboration. The justification does not simply melt into a unified concept of identity.

It would be tempting to consider the rich and complex tradition of self-narratives, autobiography, and life sagas in this context as they often hint at the possibility of coherence in a life, explanation, and justification. Many cultures have produced great self-representations in narrative form that have been preserved for us, including Augustine's *Confessions* and the life sketch of the Chinese historian Sima Qian. For our contemporary culture in the West, Goethe's coming-of-age novel *Wilhelm Meister's Apprenticeship* (1795) continues to serve as a model for this important genre. The title character runs away from home, joins a theater troupe and chases his fantasies, but constantly experiences failure. It is certainly one of the great novels of all time that challenges Wilhelm and with him the readers to assume a double perspective: looking forward *before* he makes his decisions and in retrospect considering what came to be and why. In my reading, it is also a novel of pushing away all external advice and mentoring while slowly learning to remain receptive to other voices, including his conscience. That is, many of his actions remain problematic and are in constant need of justification as they not only show flaws but actually harm most female characters in the novel.

In this context, the psychologist Jerome Bruner has emphasized the close proximity of novels and self-representations of one's life.[18] Similar ideas figure prominently in the works of the neurologist Oliver Sacks, the literary scholar Brian Boyd, and the philosophers Charles Taylor and Daniel Dennett. The psychologist Dan P. McAdams has been working for decades on the concept of the person, its development, and its narrative dimensions. He is particularly interested in "redemptive" narratives that allow individuals to turn around negative and traumatic experiences and transform them into a positive narrative of self. Narrative identity, for McAdams, consists of the condensed story of how people became what they are, or what they are about to become.[19] Narrative identity enables a life that gives people a sense of purpose; it is associated with mental health and resilience. At the collective level, groups of people are better able to coordinate their behavior and organize themselves when their members can be understood by one another on the basis of narrative identity.

However, the claim to coherence in narrative self-representation has also drawn criticism. The philosopher Galen Strawson has objected in his article "Against narrativity" that we are not guided in our actions and thoughts by a coherent self-experience, that the construct of a fixed narrative identity is actually absurd, and that there is no ethical reason why it should guide us.[20] This will lead us to a threshold of characters in narratives, which lies in fixation. I will draw a difference between justifying one-time actions and pathological self-sameness.

Identity as Pathology

The three previous activities (tracking, playing, justifying) unfold the character as an unfinished project with potential, a riddle, or a factor of uncertainty. From the standpoint of these mental operations, the character is not a closed figure and its contour is still in the process of becoming. We track, but may be led to surprising places. We play the character and coach the character into surprising decisions. And when we aim to justify, we may notice the odd discrepancies that fascinate us about the character. The character, like a narrative, can always develop and be played differently in the course of a narrative. For this enigma we have a name: such characters are *others* for us. Others are beings to whom we ascribe a life of their own, a mind which we can approach but which we cannot control and which can always surprise us. Other beings are the central challenge of our thinking and living, which we face with narrative thinking. Ludwig Wittgenstein spoke in this context of the ineluctable *other mind problem*. In playability, we presume to try out these other beings mentally, but also in this trying out we assign them at the same time a large range of possibilities for action.

There is, however, also a completely opposite approach to characters in thinking. Instead of approaching a figure as an *other*, we can also define it and suggest its identity. In this case we determine who or what the other is (or who we ourselves are). Through this fixation, the other becomes predictable. Defining and setting an identity is certainly very helpful for orientation and allows us to act quickly because we have anticipated situations. In this respect, the fixation of identity can be described as a simple heuristic, that is, as a technique by which we make predictions about the world.[21] As a result, we are not surprised by how a colleague behaves, we know what to expect from an orc in Tolkien's novels, and we can guess what a sales agent will say in a meeting. Nevertheless, the definition of an identity always includes a reductive gesture. The other (or myself) ceases to be an other and becomes a rigid figure, confined to a limited identity. The other becomes a kind of chess piece that can be used as an obstacle or sacrificed but in itself no longer plays an independent role. This effect is sometimes referred to as an identity trap.[22]

The ambivalence of this fixed identity can be seen, for example, in the victim role. The dyad of victim-perpetrator is one of the strongest patterns by means of which we sort the relationship between two figures. From a cultural-historical point of view, we have developed an astonishing sensitivity for victims—trained perhaps by millennia of religious practices and more recently several centuries of empathetic novels. We recognize pain and suffering nowadays even where contemporaries of previous centuries either did not recognize them or were willing to ignore them.[23] We are also drawn to victims today, and we often see them as morally superior figures whose suffering we empathetically witness. When we see others in a victim role, we may attach a variety of expectations to the character and mentally continue to play that role as well as blame the perpetrator. Apparently, the balance nowadays between negative emotions that we empathetically co-experience and the moral validation of us in the victim role turns out to be positive: we witness the suffering of the victims but feel positively affirmed in it, because our empathy, after all, expresses a righteous attitude.

However, there can be something highly troubling about our attention to victims and survivors. In some cases, we may not actually empathize with the victims—for example, in responding to the victim-perpetrator dyad, we may instead identify with the real or imaginary helpers of the victim, as I have argued elsewhere.[24] This move to identification with the helper shifts the moral landscape. As helpers, we can praise ourselves. We may also seek acknowledgment when we identify with helpers, because we are the good guys. If this appreciation fails to materialize, we may then develop resentment toward the perceived victim. The larger point here is that many people today are able to run the victim program in their heads as a quasi-script—and while this script certainly represents cultural progress by reducing violence, there may be some negative flip sides.

The downside of the recognition of someone as victim is the tendency for others to hold people to this role, to shrink the individual to his or her identity as victim. To be a victim is to suffer passively, to be dependent on savior figures, and to have little room to maneuver. Being confined to the role of victim is not only demoralizing but mentally deforming. It is precisely for this reason that many social movements, from the "cripple movement" to #MeToo and Black Lives Matter, rebel against this perception of victimhood and seek to give agency to those who can tell the tale. (I also do not like the term *survivor* because it still holds people to an identity shaped by the role.)

In his research on depression, Aaron Beck studied the linguistic patterns of people who tend to be depressed. In his classic studies, Beck observed an increased frequency of *cognitive distortions*, such as either-or thinking (polarization), radicalization (expressed in superlatives), and prophecies and demands ("should").[25] The cognitive distortion of interest to us here is the use of labels—expressed in linguistic forms such as "I am a . . . ," or "you are a . . ."—to mark the identity of a person. Such labels purport to establish a person's whole identity even as they reduce the individual to a state. Instead of speaking of people experiencing homelessness, people speak of "the homeless," as if this were the one defining characteristic of the human beings they refer to. Instead of saying that something did not work out, people call themselves or someone else a "loser." Instead of speaking of beliefs, people choose a label for themselves: "I am a Swiftie," "I am a feminist," and so forth. Speech patterns like labeling are a common and normal part of everyday communication, but their accumulation shows a bias. Those who label themselves or others reduce themselves or another person to a few characteristics and neglect very many others. In every act of labeling a boundary is drawn between the labeled and other people. In this respect, labeling is also exclusion and marginalization.

Such labeling, as well as most other expressions of cognitive distortion, have dramatically increased in recent decades, following a low

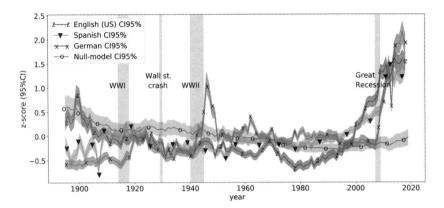

Fig. 6. Trends in the clustering of cognitive biases in the corpus of Google Books Ngram Viewer, with books in English, German, and Spanish (Bollen et al., "Historical language records," figure 3)

point around 1980, as my colleagues and I found in one study.[26] Expressions of cognitive distortions, as measured by uses in books over the past hundred years, are reaching levels that exceed even the accumulation in Germany during the Nazi era (fig. 6). Something is going wrong in the world.

Those who develop a rigid identity image of themselves, that is, who give themselves a label, may be well equipped to deal with some critical situations. Pride and stubbornness may well constitute moral strength or steadfastness. In the Baroque period, this was celebrated as a virtue, as indicated by many stoic heroes in the tragedies of the age who perish because of their convictions. But in most life situations, this firm identity means a lack of ability to compromise. Identity becomes pathology.

This pathological side of identity becomes particularly clear when we emphasize that to some degree identities always come from the outside. They are stuck on like labels or more likely put on like a corset that limits one to act according to what the corset prescribes. In this respect, identity is always discriminatory. Even seemingly positive or harmless identities are restrictive. All the worse are the many clearly discriminatory labels and categories that we have developed and cultivated in our dealings with our fellow human beings. It is a terrible experience when we are simply judged and written off based on a preconceived idea of identity. With such a pathological gesture, we are erased.

My task here is not to play language police. But from the point of view of narrative thinking, we can see how narrowed and impoverished our thinking can become when it labels other people with an identity. Narrative thinking then no longer finds a space for enriching playability. Nothing could be different anymore.

To be sure, identities and fixed roles can be a fun tool for play, playability, masquerade, and irony. In those cases, the limiting sides of identities give way to something else entirely, and we enter the sphere of the mental activities we use to create characters as others.

What Have We Learned?

The mental operations presented in this chapter give us ways to approach narrative characters. We track and play them, and then we can justify their actions. Similarly, narrative thinking makes other people accessible as narrative characters in this way, but at the same time it leaves them a sphere that remains mysterious to us. We are not butterfly hunters who stalk and pin down our prey. Rather, we keep watching them fly and flutter in various directions. We can go one step further: by means of these mental operations, we generate the characters in our mind.

By turning other people (or ourselves in autobiographies and tales about us) into narrative characters, we assume an active relationship with them. We gather information about them (track them), move them (play), and help them legitimize (justify) themselves. Narrative characters, accordingly, are beings with agency whom we simulate, whom we closely observe, and whose actions we explain after the fact, even though they also remain odd or mysterious to us. This is also what other people can become for us as narrative characters in our thinking: highly attractive enigmas or simply others, *other minds*. We are attracted to these others because the narrative strands we experience with them regularly lead to episodes of emotional reward, as we have learned in this book.

In the case of tracking, we can create an episode just from tracking a character. This is precisely the point: because we want to gain information about the character, we observe behavior and gather details. Information adds up and we can sort this information, even build arcs from beginning to end. (Satisfaction, for example, could result from this progression, when the observed beginnings lead to a deserved end.) Thus, in tracking, we register properties of the character that then lead to consequences. We track how a person behaves recklessly and then later has to pay for it.

In the case of playability, we as players are given sovereignty over the possible narrative episodes. We can mentally guide the characters in the most diverse ways and thus try out and test what could happen. Which emotions we finally end up with is largely up to us.

In the case of justification, the emotional reward for narrative thinking may lie in the happy twist. At first, the behavior of a character (or of ourselves) stumps and puzzles us. However, by explaining the behavior from within, we turn the erratic or puzzling behavior from the irrational, immoral, or unattractive to something more logical, meaningful, and coherent. Such a discovery of reasons is redemptive, therapeutic, and rewarding.

Still, in engaging characters, there is also a strong temptation to stabilize and fixate the identity of a person as if the identity would exist independently from our observations of the other. In this instance, the identity of the character comes from the outside, from the standpoint of a labeling observer. Instead of being a sphere of action and possibility, the other now exhibits predictability. Surprising actions are now perceived only as a violation of identity, not as an expression of individual idiosyncrasy. When we think this way, identity can be distorted into pathology, as others become mere pawns or chess pieces that are not affected by narrative thinking. When we think this way, we lose sight of the multiplying, non-stabilizing forces involved in narrative thinking—the subject of the following chapter.

six

MULTIVERSIONAL REALITY,
MULTIVERSIONAL NARRATIVES

Scholars of literature have the pleasure of taking the time to reflect on a narrative episode when it is complete and wonder in retrospect how it all came about. In contrast, most people do not normally use this hindsight reflection; instead, they are aware of a narrative episode while they are in the midst of it, and when it is done they rarely look back.

One of my kids comes home from school, and something is wrong. Her face isn't just the face she makes when an exam didn't go well. She starts to narrate, and a thousand thoughts pop into my head as she begins talking about her friend who seems to be in serious trouble. Narrative threads are spinning in my mind, but the central event is still missing. I imagine all sorts of things. Freeze here. This, I believe, is the standard situation in storytelling with an audience, while we are in suspense. It is this fleeting moment of many possibilities.

We take part in a sequence of events when the outcome is not yet known and we nevertheless try to guess or anticipate what might have happened or what will happen. While we are in the middle, various, even contradictory thoughts flash through our minds: something bad has happened to the friend of my daughter; the friend is moving away; the friend has behaved badly toward my daughter; or something great has happened to the friend and the shock leaves my daughter speechless. During the telling of the tale, the future is still unknown to us, even if the narrator already knows it and drops hints about what's to come that might make us feel queasy or excited. On one hand, we're always only at one point of a story, but on the other, precisely because of this uncertainty, we have more than one version of what might happen before our eyes, and we are simultaneously projecting versions of the future of the narrative.[1] This chapter will look at this tension that arises from being in the middle, in medias res. In narrative thinking, I will argue, we are always simultaneously in several versions of the same story.

I have already mentioned that we usually can only take one perspective and put ourselves in the shoes of one character at a time. Similarly, we are always mentally only at one point in time in a story. This temporality of our consciousness fundamentally distinguishes us from intelligent machines that can calculate a myriad of situations in parallel. We, in contrast, are always at one point in time and place, but we can certainly jump back and forth quickly and link past and present. Computers playing chess compute a large number of possibilities in parallel. We, on the other hand, quickly guess the best possibilities and concentrate on very few variants. We do make mistakes and often ignore good options, but we can focus well and rely on our intuition. For a surprisingly long time, this meant that excellent chess players had the upper hand by steering the game toward situations that "felt good" in the long run but were balanced in the short term. (Later generations of chess-playing machines appropriated some of the human forms of sensing and evaluating by means of machine learning and by simple preprogramming of classic situations, but this still does not mean that they "think" like humans.) Standing in one place at one point in time is a central property of our thinking and consciousness. The focus on the one moment has many consequences for our narrative co-experience. This chapter will develop the truly paradoxical consequence of this temporally confined consciousness: the multiversionality of narrative thinking that anticipates and predicts what might be coming.

Precisely because we are always at only one point in a narrative, we contemplate and experience many possible and even impossible versions of a story, in which a variety of events take place that are quite different from the story that actually occurs in the end. This is the logical consequence of being at one point: we have to invent, imagine, sense, and also reject many possible ways to continue or to explain how we got here. Because we are always at only one point, we need to keep several variants of a story in our mind at the same time.

In what follows, we will first consider predictions and anticipation and then introduce the idea of multiversionality more systematically. Building on this, we can sketch a model of the forces involved in the formation of versions. Finally, we will examine the ways in which multiversional thinking shapes us as narrative beings.

Anticipation: The Predictive Brain

Our brain is oriented toward the immediate future. Neuroscientists speak of the *predictive brain* and investigate how the brain's energy budget is geared toward predictions. Neuroscientists usually speak of the most probable next possibility, which our brain anticipates immediately, even before we are aware of it. A car overtakes us on the highway at breakneck speed, and we have barely noticed it as a shadow, yet our grip on the wheel has already tightened as we prepare for swift action and our anxiety level has already increased without our knowing why. This form of anticipation is aimed at making correct predictions regarding what we need to prepare for. Our brain tries to reduce uncertainty and to reduce and focus the so-called free energy.[2]

Classic ideas of the way our brain perceives the outside world may state that we first receive some undetermined sensory input and then focus our attention and energy on the sensory input until we can interpret it. We see a spot on a tree; we focus more and notice that it's moving. Further attention reveals it's a bird. We look closer and identify it as a robin, to use the example Andy Clark offers. This would involve a slow and laborious process that leaves us in uncertainty for a while. The predictive process jumps ahead. When it perceives a blur on the tree, it guesses that the blur is a bird. And until it's proven wrong, it operates with this assumption. Clark elaborates this idea and suggests that this process of prediction making is what constitutes human experience: we experience by predicting and adjusting.[3]

The large idea behind this process is that the brain is a prediction machine or a hypothesis-testing machine that anticipates the future. An essential part of this *predictive brain*, then, is generating predictions, or hypotheses, about what the situation is like, what comes next, and how we can profit from the situation. This involves continually adjusting what is actually occurring in feedback loops and quickly revising our predictions or hypotheses. In this case, we talk about correcting wrong predictions (*prediction errors*) and reducing surprise, which can be exhausting and dangerous. Surprise is only good if we have already anticipated it. This idea of the brain as a prediction machine is often related to the *Bayesian brain hypothesis*, that is, a brain that has to decide continuously between possibilities. The hypothesis that we generate is usually the most likely one. Errors are costly insofar as they require more attention, adjustment, and thus energy.[4]

Benjamin Hutchinson and Lisa Feldman Barrett describe the work of the predictive brain as creatively simulating the environment as if we were observing ourselves in an aquarium: "A brain is continually running an internal model of an animal's world. The model is generative, meaning that past experiences can be recombined in novel ways as they are remembered."[5] I will refer to this internal model as an aquarium. In this model or aquarium, we see ourselves as an animal and continually compare what we create as model to what is really happening. We observe the environment and the deviations of the environment from our model particularly closely; indeed, it may be precisely at such deviations that our brain really cranks up energy to home in.[6] The model or aquarium enables us at the same time to see, feel, and sense more than is possible for us at our moment in time. We can already see where the cave with the dangerous sharks might be. To do so, we use the memory of our past experiences to create prediction hypotheses of what is to come.

Some emotions can be involved as well in the prediction machine, insofar as emotions kick in more quickly than reason in grasping a situation. Fear is already breathing down our necks and we react; we already anticipate the surprise.[7] In general, we can quickly understand that the predictive brain could be a useful adaption that helps our preparedness and readiness for fight-or-flight reactions, as well as quick reactions of many kinds.

Most theories and models of the predictive brain suggest that the brain is ahead by fractions of seconds. Beyond that, the prediction errors would be too huge and costly. Still, beyond the anticipation of the immediate future, we also engage in the more distant and uncertain future, which may be concrete or still quite diffuse before our eyes. Only a few neuroscientists have so far attempted to connect ideas of the more distant future with the *predictive brain*.[8] This is, however, the starting point for our reflections on the *narrative brain*.

Multiversionality

If what I have now learned from records after my mother's death is true, then my father did not die in an accident—he was deliberately murdered by the KGB because, as a West German diplomat, he had received important documents of international significance for the Cold War. I discovered this new information while I was writing this book, and the revelations took me back to my childhood when I was ten. We were on vacation in a remote nature park by the Baltic Sea in Sweden. My father, who worked for the West German minister of defense and chancellor, had received a telegram saying that they needed him to look at secure documents, which he had to retrieve in person. He picked up the documents but never made it back to the hut where we were staying. His car was found nearby, but not him. For several days, helicopters and military ships searched for him. At the time my sister and I thought that he had gone missing on a hiking and climbing trip. His body was found in the ocean by military divers three days later, but not the documents. We thought his death was a climbing accident, and that was how the event was portrayed to everyone. Now the narrative and multiversional space of my life has changed completely and, at the same time, of course, not at all. I can imagine how our family's life would have been different if he had stayed alive. Or I can imagine how I would have developed differently as a child if I had known about the political murder, if my mother had not taken her knowledge to the grave. Who knows what kind of books I would have written then instead of this one? Would I have grown up as a person who hates Russia? I also imagine a happier life for my mother, a different childhood for my sister. A few weeks after my mother's death and the revelations that followed, Vladimir Putin launched the Russian invasion of Ukraine.

When we consider more long-term predictions and expectations, it becomes clear that we cannot simply jump toward the most likely prediction. Instead, we will also likely need to generate different and opposite variations of what could happen or could have happened. We may have a most likely or preferred version foregrounded in our mind, but even then we can also have different variations of a "back story" available in case things change. This creation of different variations and possibilities is what I call *multiversionality*.[9] Two versions are distinguished in that they contradict each other in at least one aspect and cannot both be the case at the same time. Mental multiversionality is, as I will suggest, a central characteristic of the narrative brain. Multiversionality does not only concern the future but can also include the anticipation of the past that is not yet known. In this regard, the time of the narrative and that of the plot can differ (narratologists sometimes distinguish the two by the names *histoire* and *discours*, or story and plot).

The animal in the aquarium of the predictive brain becomes the playable creature of the narrative. From the standpoint of the predictive brain, as Hutchinson and Feldman Barrett describe it, the aquarium with the animal serves as a vision of our immediate environment. It is a model that helps us anticipate what's behind the corner. However, in regard to the multiversional narrative brain, the animal in the aquarium now functions as the protagonist of our thinking and co-experiencing, and this protagonist has many options of what it could do or what could happen to it. Those who simply follow a story as if it were a track of a-b-c-d do not think narratively. Narrative becomes a form of thinking when we anticipate contradicting possibilities.[10]

Suspense is a particularly clear case of multidimensionality, so we can use suspense to explain multidimensionality. When something becomes exciting, we usually expect some negative course, but we pin this expectation against the hope that everything will turn out differently. Scientific studies on suspense usually distinguish between versions of fear and hope.[11] Suspense keeps us on the edge of our seat as we wait and hope with excitement for the better of the possible endings. Suspense involves a postponement of the uncertainty's resolution, and, in that postponement, anticipation of various possibilities intensifies.[12] In this process, the expected negative version may stand concretely before our eyes or lurk in our mind, while the positive alternative appears more vaguely as hope. Works of fiction train us to make this distinction between the feared and hoped-for versions. And in works of fiction, the feared version, which is usually the more probable outcome in the real world, is usually the one that does not occur after all because an improbable, but hoped-for, turn of events intervenes. We know this from fairy tales and in fact most stories in general. Of course, rationally speaking it's not likely a prince will rescue Cinderella from her misfortune, unless of course we're in the world of a fairy tale. In works of fiction, therefore, we can usually experience suspense as something quite positive and pleasurable, because the improbable hope is usually fulfilled.

Postponement also contributes greatly to the enjoyment of suspense in fiction because it involves an intensification of perception. That is, the mental generation of an actually improbable version is rewarded threefold: first, this version is positive; then when it occurs it thereby sanctions our prediction; and finally, we associate tension with a positive feeling because uncertainty increases the sense of presence.[13] Precisely because we are compelled to consider other points of time, we also spend longer with this present moment and thereby intensify it. The one point in time becomes loaded with tension.

An interesting phenomenon related to these topics is known as the "paradox of suspense," which is also related to the so-called spoiler studies.[14] Under certain circumstances, a narrative is perceived as more positive if the outcome is already known to us. People like to watch some films over and over again, although—or probably because— they already know the outcome. Something similar is also known about repetitions of musical works. On the second hearing, the musical work is regularly rated as better and of higher quality than the first.[15] Surprise is best when we already know what's happening. In the model of the fish in the aquarium we are both the spectator and the fish: as spectator we see what is coming ahead of time and correctly predict it, while we as fish remain surprised.

In real life, the tension is different. We look forward to the outcome of a difficult surgery with fear or stress. Hope and fear go back and forth here as well, but we cannot gain much enjoyment from uncertainty and tension. When I have reason to wonder whether a relative was involved in an accident announced on the news or, less dramatically, when I have to make a tight plane connection, this is not a happy state. In general, surprise tends to be valued negatively in many real-life situations, which is why the predictive brain, according to common understanding, aims at reducing surprise and uncertainty. Narrative thinking transforms something that is evaluated as negative in normal life (uncertainty) into a pleasant perception that is associated with many positive effects. Regarding suspense, narratives thus have the therapeutic effect of relaxing and reducing stress. Suspense is also associated with increased immersion in stories.[16]

Now we can ask more precisely how narrative thought (and works of fiction in the narrower sense) perform a positive reshaping of perceptions that are normally strenuous and judged as negative. In earlier chapters, we have sketched the following answer to this question: narrative episodes reward us with emotions; already at the beginning of a narrative we expect or hope for such a narrative emotion as a reward. Now we can add a second hypothesis that emerges from the image of the aquarium. In narrative thinking, we are on one hand in the midst of the narrative flow, but on the other hand we are also in a purely mental space, insofar as the aquarium is, after all, only a simulation or model of us and our environment. We are offline.

Being offline presents us with some interesting possibilities. It allows us the possibility that two or more conceivable versions can coexist in our mind even though they cannot be the case at the same time. Unlike the standard situation of the predictive brain anticipating the next immediate perception, this plurality of versions also means that we can be aware that these mental representations or simulations are just that: representations, mental ideas, mere versions, figments, possibilities to which no reality corresponds (yet). We are aware that these versions contrast with our concrete reality. We have a metacognitive awareness that these versions are imaginary so far. This awareness is relieving. The narrative brain has a tremendous advantage over the predictive brain in that it can afford fictionality. The predictive brain has to react quickly and its model has to be ready for quick updates, implementation, and reactivity. Awareness of fictionality would only interfere here. In the narrative brain, in contrast, mechanisms similar to those in the predictive brain can be used without resulting in quick decisions. This gives pleasure. Enjoyment is provided by the intensification of the present in dwelling on the possibilities and the knowledge that our mental representations are not serious.

The narrative brain allows us to practice prediction voluntarily and with pleasure. This is great from an evolutionary point of view. We simulate ourselves as fish in an aquarium and explore its playability—think about what we could do there, prepare ourselves for what could come—and we enjoy these cognitive performances because everything is not (yet) real and because they can deepen our sense of presence as we perceive this presence as urgent.

Accordingly, one could propose here the thesis that the predictive brain and the narrative brain go hand in hand and that the latter is an extension of the former. The predictive brain switches to allow plurality in some cases. Narratives arise in our minds when rapid anticipation turns into dwelling on possibilities. Narrative thinking allows itself to spin out the various possibilities and also to mentally test competing versions. This allows us to simulate, play out, and then plan future actions. It prepares us for the always uncertain future and it provides pleasure in the process. *Prediction, playability, planning*, and *pleasure* combine.

Model of Multiversional Thinking

There are three different ways in which we make narrative predictions and create versions of a narrative. First, we set up a framework of probability in terms of what we think is possible (this we call *constrained expectation*). Second, we generate hopes and project wishful images of how everything should turn out (*preference projection*). And third, we make concrete assumptions about what might happen in the near future (*predictive extrapolation*) and what might explain the current situation (*interpretive extrapolation*). What follows gives a bit more context.

1. The *constrained expectation* can be fed by a number of extra-narrative and para-textual elements. For example, if we know who is telling us something, we already have expectations about what is to come. We bring a lot of prior knowledge with us here. When my

daughter tells me something after school, there are a few expected versions. Storming the teacher's room, for example, is not probable but still falls into the realm of the possible. In the case of textual and fictional forms, the genre of the text can tell us within which realm the story will take place. A book published by Yale University Press comes with certain expectations, and we would not usually expect to find a how-to guide for dating there. Such expectations give us the outer boundaries between which the events of a story are likely to take place. In some cases, these boundaries and expectations can already provide us with a schematic idea of a possible course of events. As we go through a story, the boundaries and expectations can shift.

2. *Preference projection* sets an overall goal for a story or episode. Daenerys Targaryen is to ascend the Iron Throne. The Dalai Lama shall escape despite the Chinese occupation of Tibet. My mother will find her happiness again. The Eagles should have another championship game. The extinction of the whooping crane should be prevented. Such preference projections do not necessarily guide our concrete ideas about what should happen in the next moments but represent a distant goal to which we orient ourselves. Using this distant goal as a guide, we can assess the specific events that will occur and the extent to which they are consistent with our desire. Without such an "emotional investment" a narrative would mostly be boring—and perhaps no longer a narrative at all.[17] Many concrete preference projections can be congruent with the narrative emotions that reward us at the end, such as triumph, love, or even deserved punishment.

3. *Predictive extrapolation* continually generates versions that emerge from the concrete material of a story. "What did the teacher say in response?" We use our general knowledge about probabilities, about how events are related and how people behave, to create inferences.[18] These assumptions are constantly renewed and adapted to the situation. These inferences and assumptions can also be interpretive or retrospective (*interpretive extrapolations*). That means we can make guesses about what happened earlier. I see from his face that his date last night was not good. What happened earlier might explain to us how a character arrived at a particular story situation.[19] Interpretive

extrapolations lead to new ideas and models of the past and present (which, of course, then lead to new future possibilities). You don't have to be a detective to do this.

All three forms of prediction have in common that they start from the one moment in time that we occupy at any given point in a story. As story creators and audience of narratives, we are always here in the moment. And from this point of presence, we create predictions. Taken together, these three forms of narrative predictions present us with a picture of narrative experience in narration, or narrative development from one point in time to the next (fig. 7). At the later point in time (seen in the graph as a shifting star), many of the developments previously thought possible have either occurred (A) or been discarded (B), and new possibilities have emerged (C). The projection of preference has shifted because the earlier preference is now outside the range of the adjusted *constrained expectation* ("Okay, then she'll marry another prince, or she'll take the throne all by herself").

Even a discarded version can retain some mental presence in narrative thinking. In contrast, in the predictive brain, we can assume that the multitude of very quickly updated possibilities disappear again immediately, usually leaving no trace or memory. In narrative thinking, however, preferences often persist even when they are factually impossible. In works of fiction, of course, we are trained to still hope for the impossible, because in many works those believed dead come back. But we also carry this narrative wishful thinking into our real lives. I continued to fantasize about my father's return for decades after his death.

Nevertheless, many versions disappear, become impossible, or simply do not occur; however, this happens slowly and they do not vanish entirely. We can distinguish between different forms of version disappearance—for example, slow *fading*, when there is little evidence for the continued possibility of a version; preference change or *spurning*, when the preference projection turns out to be a mistake ("that guy is a sociopath"), but even a spurned love can flare up again; and finally *foreclosing*, when a version becomes factually impossible.

In regard to novels, Karin Kukkonen has done great work on the

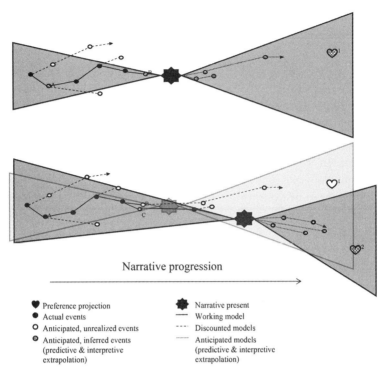

Narrative progression

♥ Preference projection
● Actual events
○ Anticipated, unrealized events
◎ Anticipated, inferred events
  (predictive & interpretive
  extrapolation)

✴ Narrative present
— Working model
---- Discounted models
······· Anticipated models
  (predictive & interpretive
  extrapolation)

Fig. 7. The formation of mental versions and expectations in the course
of a narration. The recipient is located at one point in the narration (black
star), from which the possibilities of development are formed in each
concrete way. The gray triangles mark the frame of the *constrained
expectation*. The lower figure shows a later point in time where
what is possible—but also what is hoped for—has shifted.
(First published in Benjamin Hiskes et al., "Multiversionality")

consideration of emerging probabilities. These probabilities concerning what might come are close to the versions of our model. Kukkonen differentiates three levels of predictions in novels:

1. Probability at the level of plot.
2. Probability at the level of perceived reliability of the narrator and
   narrative tone.

3. Probabilities derived from integrative interpretations of major patterns and the overall text.[20]

Narrative Thinking

In narrative thinking we duplicate our real world. We take our starting point from concrete situations in which the protagonists of narratives find themselves. These situations incorporate many constraints that we comprehend, such as who does what to whom and why, when, and where.[21] We also tend to simplify narrative situations by means of stereotypes, schemes, and scripts that are familiar to us. The tendency to quickly categorize guides our thinking. Once found, explanations, justifications, and patterns condense and corroborate.

And yet narrative thinking also allows us to detach and disconnect from the real world. We simulate the real world in our thinking, create a small "aquarium" in which we observe ourselves or the heroes of the story. In this space, we can push our heroes back and forth or let the aquarium spill over sometimes to test what else is possible. The aquarium is a mental world in which we can experience extremes that we would not wish for in the real world.

With multiversional narrative thinking, every pessimist can also try out the positive dimensions of life on a daily basis. People who expect something negative likely still harbor the hope for the positive; they just mask it better. Even people who are stigmatized by the strongest stereotypes can break out of their confines by means of narrative thinking. Alternative realities can occur to us and be explored and co-experienced. We can experience stories from the most diverse perspectives and try out even the less probable variants of possibilities. While we may also harden and close our minds, even in these cases the perspectives of all parties are accessible to us, and secretly we try out the perspectives of others and versions that are different from those we know.[22] This does not make us ethically better people, but people who live richer lives because we mentally inhabit many worlds. We can experience the most diverse possibilities as versions. Narrative simulation is thinking and experiencing in the plural.

My proposal to consider multiversionality as an essential aspect of narrative thinking will baffle some of my readers. In narratology, for example, this notion or similar ideas have not yet caught on. In fact, when I first published papers and chapters with this idea a decade ago, I could not find even a ready name or concept to explain the idea. The odd term *multiversionality* had been used just a few times, and mostly in regard to medieval manuscripts that existed in more than one copy but with differences. (To be sure, I do find kindred ideas in works on suspense and discourse processing.)

Many scholars and scientists may be reluctant to consider multiversionality because traditional oral narratives, like works of fiction, are precisely "linear" or "one-versional," starting with the linear arrangement of writing. We would not want to read multilinear texts, or only if they are composed by amazing artists and hang nicely framed in a museum. Another reason for reluctance might be that it seems hard to imagine how the messy process of making sense of stories happens in our mind. In this respect, the proposal of multiversionality might appear almost too clean and neat. It's true that one should not underestimate the creativity in the reception of narratives. The main reason, however, for the reluctance is that many disciplines of reading stories consider stories ex post from the point of completion. Instead, I think it is vital to consider stories from their middle, from the perspective of being in medias res, before all is known and complete.

Perhaps our reading habits will change along with our recording media, however. Video games have already gained a firm place in the daily routine of many people. There, quick decisions take place that lead to different versions of the plot as each player experiences a somewhat different story. Interactive fiction is also rapidly spreading and gaining popularity. Readers or players of these works try out numerous versions of a story in order to compare them with each other. In the immensely successful *Detroit: Become Human*, for example, readers are provided with a map for every decision they have to make, showing them how many other readers have made the same decision at that point. Apparently, this motivates many readers to return to these points to try out different versions.[23] Following the success of *Black Mirror: Bandersnatch* (2018), Netflix has also invested heavily in the expansion of interactive forms that involve viewers in the development of the plot and transfer decisions to them.

This multiversionality also manifests itself in other new and old genres. In every tragedy, there is always the hope that everything could turn out differently. And in fan fiction on the internet, the followers of a story spin it further or have their heroes appear in the real world as well. During Comic-Con festivals, entire cities are transformed into a carnival, with fans dressing up like their favorite characters. In a conference hotel, I was recently accompanied by six orcs in an elevator before a lecture on the ideas of this book.

What Have We Learned?

Earlier in this book, we learned about principles that govern narrative thinking, such as the tendency for us to separate actions into episodes with a beginning, middle, and end—narrative episodes that drive us toward rewarding emotions but also allow us to retreat at the end. In this chapter we have added a new and central idea to narrative thinking and narrative co-experience, namely, that it involves the mental production of multiple possibilities and mental versions of the story.

In narrative thinking, we are always in the middle of a story: what is yet to come is uncertain but not arbitrary. We are focused on anticipating, guessing what is to come, and hoping for the best. In narrative thinking, therefore, there is always more than one version of what is happening or could happen that we are aware of. We are mentally in spaces of possibility and thus in a multidimensional space. Not all versions are concretely "in front of our eyes" step by step like a movie, but we know that these possibilities of sequences exist and we already have an idea of what might await us. In this chapter we considered the phenomenon of suspense, but we could just as easily have taken detective stories, narrative thought experiments, or the phenomenon of surprise as the starting point for our considerations. Mental multidimensionality appears in every narrative. If someone does not experience at least some mental multiversionality while hearing or reading a story or the latest gossip, I would suggest this person did not engage in narrative thinking during the process.

To understand the differences between narrative and non-narrative thinking, we compared multiversional narrative thinking to the neuroscientific notion of the predictive brain, that is, a brain that continuously and very quickly makes predictions about the environment, eliminating uncertainty, surprise, and false predictions. The narrative multiversional brain extends the predictive brain into more long-term predictions and into a space where contradictory predictions do not have to be immediately reduced to one and thus can be safely entertained. In narrative thinking we award space to imagination or fictionality, decoupled from the reality principle.

seven

EVOLUTION OF THE NARRATIVE BRAIN: THE STAGE AS THE BIRTHPLACE OF THE MOBILITY OF CONSCIOUSNESS

There is one more decisive and mysterious achievement of our brain left for us to discuss: not only do we mentally reconstruct stories we hear or witness, but we put ourselves into this story as if it were occurring here and now. When we imagine a story or hear it, we are *in* the story. The story world—Harry Potter's Hogwarts, for example— is the center of our conscious experience; we have an almost bodily perception of the narrative, that is, *embodied cognition.*

Narratives, by virtue of their episodes, are great for remembering and planning events. The macro and micro episodes that allow us to encode action make the complex events manageable by organizing them with a before and after. We can track, play, and justify the characters, and we can adjust to multiple versions of the same story. And because of its emotional rewards, we enjoy narrative thinking as well. However, all of these cognitive activities hinge on our ability to experience or co-experience the story as if it's *our* story, taking place in our present moment and concerning our bodies. Narratives draw us in, transport us into their space, and thus immerse us.

Humans are characterized by the *mobility of consciousness,* or *mobile consciousness,* which allows us to put ourselves in all sorts of different places and situations. Because of this mobile consciousness, we have broken the boundaries of the present and live not in a single present but in many worlds and times. We have not ceased to be animals, but we can mentally detach ourselves from our real-life situation and think ourselves into other situations, other worlds, other times, and other beings. In fact, we not only "think" ourselves into these other presences but also understand, experience, and feel them with our entire bodies. With every narration, we immerse ourselves in a multidimensional virtual world, without the need for special glasses and headwear. Large parts of our brain are engaged in this process; in fact, they are the very parts that would be active if we were in these situations ourselves.[1]

There are various words for this phenomenon—for example, psychologists speak of transportation and narratologists of immersion.[2] In the field of virtual reality, people often speak of simulation. These terms also describe the effect that narrative worlds can have on us when they draw us into the gossip of friends or into a fictional world. I propose to use the notion of mobility of consciousness for this phenomenon to describe the activity of our mind to find places of residence for itself. In this chapter, we will focus on this capacity. What does it say about us that we possess this faculty? How did we learn it? What constitutes our mobile consciousness?

Consciousness is certainly a difficult concept. Those thinkers who undertake the study of consciousness are always under suspicion of having lost their marbles. I'll have to assume this fate will also befall me. I can't claim to know what consciousness is.[3] But at the same time I hold it to be true that narrative thinking puts us mentally into nonphysically present situations. Hence, I am forced to consider consciousness from that perspective. That which is moved, displaced, and relocated by means of narratives, but which, paradoxically, at the same time constitutes the very center of our perception, I call consciousness.

Specifically, mobility of consciousness concerns the human ability to move the center of our perception into diverse situations. This means we can withdraw our consciousness from our physically present moment and immerse ourselves into a past or imaginary scene. If we first understand consciousness as the sensory perception of the environment in the present, this mobility of consciousness adds a dynamic element to perception: we perceive a presence, but it need not be our physically present one.[4] In some respects, this mobility of consciousness is related to imagination, but with specific differences: mobility of consciousness does not require invention, as we may receive the world we enter from someone else. Unlike imagination, it also invokes a sense of presence in one moment at a time.[5] Most importantly, then, mobility of consciousness enables us to experience someone else's world: we can co-experience what they have relayed to us. Even when we re-experience one of our past memories, we slip into our own shoes again thanks to this mobility of consciousness, and we are, to some degree, someone else. In short, mobility of consciousness also plays a key role in some forms of empathy, namely, forms that rely on the co-experience of the situations of others.[6]

The question of this chapter is how this mobility of consciousness could have been developed. The subject of discussion is more precisely the question of which *cultural evolution* could have promoted this ability. I propose to locate this cultural evolution in the development of the "stage," understood in a broad sense as a site of gathering and performance in the presence of others. We learn with and from others that a performance stages a story we can mentally enter as if we were there. This is how we learn to co-experience stories.

Gatherings, and perhaps stages, may have already shaped our group behavior for several hundred thousand years.[7] "Stage" in this sense is not so much a physical place, but rather the collective event in which people become performers. Such stages may have existed long before the architecture of theaters, and their origins may be linked to the telling of myths or the recapitulation of past events. The thesis of the chapter, then, is that the cultural institution of the stage and performance has trained us in the narrative processes that develop mobility of consciousness and, furthermore, that our consciousness is shaped by this mobility and thus narrativity.

Stages and performances in front of others provide a collective focus for our attention and offer the staged events up for collective co-experience. In this respect, stages and collective empathy are structurally linked. Collective empathy involves a group of beings who empathize with a person or situation. Collective viewing and observation, I will suggest later, may have played a role for cultivating empathy by synchronizing the behavior of the audience via collective rhythm, breathing, and heart rates.[8] For the following considerations, it is not vital whether this act of collective empathy results in a strengthening of the collective—which, however, seems likely.[9]

By means of this mobility of consciousness we have freed ourselves from the confinement of our physical presence. We are not limited to a singular presence or a single existence. Whether old or young, athletes or couch potatoes, rich or poor, we can all roam Middle Earth, be thrilled by Patrick Mahomes or enchanted by Taylor Swift, transport ourselves to an erotic fantasy, or relive a memory. They say fame and fortune don't make you happy, and according to many studies, that's how it actually seems to be. All of us can dream ourselves into the situation of a Cinderella or a Harry Potter. We can, usually, choose the memories we want to relive. We participate in other people's lives and conversely know that others participate in our lives—we are not alone. We deepen our own lives by revisiting and reliving certain situations over and over again. Or we break out of the narrow confines of our lives as gamers and battle in virtual worlds. We can delight in our triumphs but even enjoy our failures. In short, our mobility of consciousness makes us more than we are here and now.

Nevertheless, we are not completely unbounded. Our consciousness is temporal—that is, we are mentally always only at one moment in time and only in one situation. Even when we solve a mathematics problem, we are always only at one point in the calculation at any given moment. Our so-called working memory is also narrowly limited.[10] In general, we can jump back and forth quickly, but this singularity of one moment at a time remains formative for consciousness (with the oscillation as in Kant's *Critique of Judgment* as limit). Despite mobility, our consciousness remains situationally bound and, as I would emphasize in this context, entangled in narratives.[11] What our consciousness processes is also likely to be relatively confined in time. We probably only perceive a few essential features of a situation or a respective present, which give us the impression of a presence, and we add the rest only when needed.[12]

With these short sketches, we have circumscribed two aspects of consciousness. First, consciousness is temporally confined; it is concentrated on a single point in time and space. Second, consciousness is mobile—that is, the one point in time and space is not completely predetermined by our physical and sensory presence, but can also be mentally directed by certain procedures to real or fictional times and situations, allowing us to co-experience them mentally. These mobility-enhancing procedures include observation, imagination, and transportation in fiction and narrative, memory, and various forms of artistic communication.[13] We will look at some of these procedures in this chapter and consider them as cultural techniques.

Based on the two properties of consciousness—its limitations to one point in time and its mobility—we can identify some features of consciousness. For our purposes, we leave aside other possible forms and states of consciousness, such as generic consciousness (for example, what it is like to be a bat?) or meta-consciousness that reflects on its state (I know that I am sitting here writing).[14] This means that we locate consciousness in the vicinity of episodic memory and narrative thought, similar to some of the descriptions of consciousness that Douglas Hofstadter and Daniel Dennett have offered.

We might begin by asking whether these two properties *essentially* belong to human consciousness. If that is the case, we could no longer simply assume that mobility is a later or individually developed evolutionary addition to consciousness. Rather, we would have to assume that consciousness could *only* be developed with this inherent property and should therefore be understood as an empathetic consciousness. Consciousness would therefore not be attributed to a thinker with his or her insular ego. Consciousness could only exist in the plural; for every individual with consciousness, there would have to be others to relate to. Such individual consciousness would have to invent other conscious beings, if they did not exist, to slip into their world for co-experience. Without mobility of consciousness, there would be no human consciousness. While this goes beyond our matter at hand and what we can even begin to develop here, it could at least be suggested that a consciousness without this mobility would be fundamentally different and actually hard to imagine. Only a few odd and lonely philosophers might be up to the task to imagine a solipsistic consciousness, one that focused on one being alone without mobility and without co-experiencing.

In place of this solipsistic consciousness, I envision a human consciousness that evolved to incorporate mobility. What skills did humans have to acquire biologically, and what did we have to practice culturally, to develop this mobility of consciousness? How did we have to reshape our social and cultural living environment to promote the development of mobility of consciousness?

There are further questions. The mental transport that is made possible by the mobility of consciousness does not apply to objects, but to another being to which we impute a consciousness (if we focus on an object, we end up treating it as if it were conscious). Other people and beings are a tremendous mystery to us, perhaps *the* mystery of our lives. What is going on in others and what moves them, we can never determine for sure from the outside.[15] Our mobile consciousness allows us to experience this other being as a self, and we transform our third-person or second-person perspective (they, she, he, it, you) into a first-person perspective (I). Earlier we talked about the playability of others. How do we do this?

One proposition of this chapter is that our narrative thinking and mobile consciousness could mutually strengthen each other because they were practiced and learned as a collective form of empathy. To develop this line of thought, we will start with the collective dimension of empathy, before we turn to its connection to narrative thinking.

Today we think of empathy and mobility of consciousness as behaviors that we perform as individuals, like talking to a friend or watching a TV series, but this is probably not how our ancestors acquired this ability. I suggest that some individuals started to act out events in front of others and that the group as a whole got involved, as spectators or as co-performers, perhaps as in the sense of a dance. In a performance, something is made present and embodied before spectators.[16] The educational impact of the theatrical performance is immense, because the spectator can be directed to what is being represented.

The central question here is how the other, that is, the performer, could become the projection figure of the observer's mental, mobile identity. To address this question, we will look at the development of collective empathy through three elements of performance: joint attention/joint performance, the performer, and the cultivation of receptivity. In general terms, these three elements correspond to the three cognitive actions by which we approach characters in fiction (as described in Chapter 6): tracking, playing, and explaining/justifying.

Joint Attention, Joint Performance

Joint attention is a prerequisite for the synchronization of behavior. Staging a performance requires joint attention by the audience. The evolutionary biologist Michael Tomasello and his colleagues view joint attention as a core developmental accomplishment that has enabled communication. Over many decades, they have conducted comparative studies with chimpanzees and humans. Although young chimpanzees outperform human children early on in terms of motor skills, it turns out that human children develop a better willingness to cooperate early on; they help each other, communicate clearly, and develop more social skills overall. (But chimpanzees also show remarkable abilities. In competitive situations, for example, chimpanzees are apparently capable of assessing what other individuals see and what they do not, thus exhibiting at least a rudimentary theory of mind.) Tomasello and his colleagues identify *joint attention* as one of the decisive abilities of human children.[17] Groups of people regularly direct their attention to the same object. This coordination of attention can, of course, simply come from a particularly exciting or attractive object that fascinates everyone. But people can also track where other individuals are looking and follow their attention. The development of the white sclera in the eye, which is largely absent in non-human primates, promotes this ability; accordingly, it is thought that this white in the eye has evolved adaptively to promote tracking of gaze direction.[18] A person who tracks the eye movements of another receives two kinds of information: she observes an object, and she receives social information about the other's attention.

We do not just follow where someone else is looking; we can also purposefully get others to follow our attention. That is, we can point. One can point with the eyes and head movements, one can use an arm as a pointing arrow, and one can point to an object or circumstance by means of symbolic codes, such as language. Tomasello has suggested that this ability to direct attention by pointing is a foundation of language development, prior to and more fundamental than communication by sounds.[19]

Joint attention is indeed a complex phenomenon. Building on Tomasello's work, we can distinguish the following forms:

1. Attending to an object by several individuals (*object knowledge*).
2. Feeling of a commonality as others are also fixated on this object (*we-feeling*).

3. Knowing that the other is fixated on this object (*subject knowledge*).
4. Directing the other to an object, perhaps by pointing (*ability for manipulation*).
5. Understanding that the other wanted to draw my attention to this object (*understanding of manipulation*).

These different forms of knowledge can be activated to different extents in concrete cases. A leopard certainly attracts the attention of many individuals (1). When a group of individuals starts to become restless, this restlessness may infect others, alerting them to the approaching danger. The group coordinates, reacting as a group—for example, by choosing the same escape route (2). However, the concrete knowledge that one individual is fixated on an object can also be helpful to another in many ways; for example, an individual could take advantage of the distraction of the others. While everyone is looking at the leopard from a safe place, secret snacking can take place (3 and 4). The knowledge that another individual wants to draw my attention to something and manipulate me (5) already presupposes a complex understanding of communication. Its content here involves knowledge about the intentions and internal states of one another. This is where theory of mind or cognitive empathy begins.

Curiously, the scenario of the leopard could also take a different form. It could be that some group members observed a leopard earlier and they are still nervous and agitated about it later when they rejoin the larger group. Perhaps they start to act out the movements of the leopard, consciously or not—that is, they use *playability* to become a leopard. This could be quite puzzling to others. However, the others might also participate in this spontaneous event of imitation by also imitating, thereby joining the stage, becoming performers, dancers. This too could be an origin of co-experience. Co-experience might start with observing or with joining. Either way, when others stage an event for us—when they display a past or imagined experience for us and make it available—co-experience can begin. This means we need to consider the performer.

The Performer

Performance requires a performer, someone who draws attention to himself to perform for others. Attention alone is not enough for performance. An individual who assumes a leadership role has already gained joint attention to shape the behavior of others—for example, to initiate the departure of a group in a particular direction. The leader rises, others follow. This leadership, however, is not in itself a performance, as the leader does not care what is going on in the minds of others as long as they follow. What is missing here is the active creation of an illusion.

Performers play a different person, a different being, or themselves in a past or imaginary situation. That means they act something out that doesn't fit the current situation they are in. Instead, they create an illusory situation by their acting and playing. Playing and playability initiate the process.

Once such a situation has been created by means of performance, others can react to the creation, assuming that the performer was successful in her play. Others could now *be deceived* and take the situation for real. If an individual plays being scared, spectators could assume that there's something dangerous out there. Others could also *join in* and mimetically play along, for example, by also moving as if they were a leopard. After all, performance and play are contagious. Finally, others could also *observe* the performance from a distance and co-experience it.

In the case of *deception*, the performer may have purposefully created an idea in the others' heads with the hope that they would take it as real. Whoever knows how to direct the attention of others and can give it content is also able to deceive them. A magician knows he can direct everyone's attention in a certain direction—for example, to one of his hands—while he makes something disappear with the other. This misdirection actively creates the imaginary content for the spectators, which he knows hides something and does not correspond to the facts (for example: the cloth is not in the hand). Such manipulation can then lead to the creation of a story in the mind of another, which the performer knows to be more or less different from alternative possible versions.[20]

To be sure, the existence of a deception does not yet mean that one actively and willingly deceives the other. An example from behavioral biology can explain this. A baboon that mates outside the tolerated harem structure can expect aggressive behavior from the high-ranking male. Nonetheless, non-tolerated mating does occur. When this happens, the baboons exhibit behavior that we would interpret as deceptive behavior in humans. For example, a pair suppresses the typical copulatory sounds and seeks out hidden locations; individual baboons have also been observed covering their swollen genitals.[21] But whether this behavior involves real awareness of the deception, and thus manipulation knowledge, remains controversial. After all, it could simply be that individuals have learned from experience that this behavior (such as mating with low-ranking males) succeeds only under certain conditions, such as hiding. Behavioral biologists are appropriately cautious about using the term *tactical deception*. The more plausible assumption is that an individual has simply learned by trial and error to avoid problems such as aggression by the dominant male. This does not yet presuppose any knowledge of the imagined content of the deceived. The baboon succeeds, but whether the baboon has willful knowledge of manipulation is unclear.

In the case of *joining in*, the group would be prompted by the performers. Someone starts acting like an animal and all can imitate the movements. The playability of the performer takes over. All it takes is the start, the initiator, the "lead." As we all know, there are few activities as satisfying and enjoyable as being in sync with others, coordinated by some form of rhythm or sequencing of motion.[22] Physical movement sets many cognitive and affective processes in motion. Although it probably does not in itself lead to the co-experience of the situation of another, it might provide a strong basis for joint experiencing.

In the next section, we will get to the conditions that train the *observation* of the performance of another. To some degree, all playing creates an illusion for others.

The all-connecting element of the creation of an illusion is the role of the one who generates these illusions. Such illusions can be recalling recent events, sharing the history of the clan, outright deceptions, excuse-making to avoid blame, a pantomime to teach successful hunting, a narrative account of what the future might bring, or an elaborate opera by Mozart. Accordingly, there are many reasons that inspire a performer, from the pleasure of imitation to the intent to deceive or distract, to advertise oneself or glorify someone else, to disseminate information or preserve a legacy, and so forth.

The Cultivation of Receptivity

A performance needs an audience, because without observers it would not take place. The performance is there for an observer. The success of what is performed is determined by what is going on in the mind of the observer. In a classic study of theater, Richard Schechner argues that the theater of transportation, which conveys the audience into other worlds, and the "ritual of transformation," which changes the participants, are related in many ways, insofar as the audience plays an essential role in the realization of the performance. In many societies, therefore, the spectators are also the privileged ones. The horizon of the performance is the mental event of the audience.[23]

The spectators who perceive an action performed by others need to explain or justify this behavior. If they cannot manage to do so, the actions of the performers are likely nonsensical for them. Only if they transpose the performance into a performance *of* something does it succeed. To explain why the actor is walking weirdly, the spectators must explain it as the walking of a leopard.

Such receptivity needs to be trained and cultivated. Friedrich Nietzsche devoted significant parts of his thinking to understanding how the training of the receptive mind could lie at the foundation of culture. According to Nietzsche, this involved submissiveness toward the dominant speaker or performer, which he explicated in *Beyond Good and Evil* (1886) and *On the Genealogy of Morality* (1887). Even if one does not follow Nietzsche in this respect, this idea brings to mind that receptivity requires learning and is guided by expectations. Kids routinely get hushed to be quiet when others are speaking.

Apparently, being quiet, even prior to receptiveness, is not a natural act for them.

The dyad of actor and audience, speaker and listener, or performer and observer is one of the most basic guiding ideas of our civilization for communication. The coordination of performer and audience requires a range of cultural mechanisms. In everyday communication, the repertoire of cultural mechanisms includes intervals of back and forth between talking and listening; it comprises many encouraging signals by the audience, such as visible attention, synchronized movements, or nodding; and it can involve the asking of questions. In modern theater, this includes the drawing and falling of the curtain, which marks the beginning and end of a performance, but also clapping and collective silence. In smaller groups, it is especially likely to include the spilling over of emotions, such as excitement or indignation over the portrayed circumstances, as well as side-taking; this can occur in large groups as well.[24] Ritualization and repetition are also factors in learning how to be part of a receptive audience. We learn that an event can be portrayed repeatedly, and this reiteration indicates that something else is being referred to in the present portrayal.

Performers learn to gauge the success of their performance during the staging and make changes to adjust it. If, for example, hunters mimetically stage a particularly effective hunting technique for their audience, they can measure the success of their performance by the excitement of the audience and, if this fails to materialize, start again.[25]

Communication about events that are spatiotemporally absent or non-existent is a tremendous feat. The leap from our closest relatives, chimpanzees, to us humans is striking in terms of communicating past or imagined events. Although non-human apes probably also use referential sounds, they probably do not succeed in representing past events.[26] We will not speculate here on which biological differences are responsible for this. Rather, we will discuss the conditions under which the mobility of consciousness could be culturally acquired, whether by chimpanzees, Neanderthals, *Homo sapiens*, or artificial intelligence. We do not know what chimpanzees or other non-human animals could accomplish if they underwent enhanced cultural development, such as the development of language.[27]

For us modern people, communication of past, absent, and even imaginary events is an easy feat.[28] We have language that can simply state what happened. When someone tells us something, many people "see" the absent event before their eyes and experience it. Even people with aphantasia (who do not "see" absent events) still relate to the characters and emotions. Narratives encode the sequences to which we then may add different forms of sensual, emotional, or intellectual embellishments.

The audience and listeners succeed in perceiving the non-present event as present. For this, the mimetic and theatrical representation is certainly well suited, since it contains a clear physical presence. The audience simply needs to ignore or forget some aspects of its observation, such as the identity of the actors they know, in order to understand that they embody others in the context of this performance (or are revealing a past moment from their lives). Casting becomes a cognitive performance. At the same time, they have to add elements that fit the situation to make it appear more cohesive.[29] That means the audience must both turn away from some aspects of the performance (the identity of the performer, the concrete surroundings of the staging, the present moment) and supply the missing elements (sensory experience, context, and the like). Only then, the leopard in the steppe can appear in their minds.

The spectators also need to leave their physical bodies behind to embrace sensory perceptions that are only in their imagination. Only those who know themselves to be safe will do this. Collective reception could make this easier. If others are also watching, there is a high probability that one is safe. In this respect, the collectivity of reception is not only a means of attuning and coordinating the audience, but also—at the same time and even more rudimentarily—the condition for drawing attention away from the environment and concentrating attention.

We now come back to the site of the performance, the place of the staging. Obviously, performances can arise spontaneously at any place. When we tell the latest gossip, we do not have to wait until we are in the theater. In the context of our argument, however, if we assume that performance for others is a form of improbable communication that must be learned, we can now point to the need to cultivate and optimally attune actor and audience to each other. This occurs when the stage as an institution becomes ritualized. A stage in this sense is the site where something is performed. This can be the cave of the Stone Age people, the common sleeping place or the meeting place where objects of practical or religious significance are remembered or performed, and, later, the stage of theater. Perhaps theater emerged from ritualized fights or from collective music or dance events (as already imagined by Friedrich Nietzsche in *The Birth of Tragedy*). Music and dance with their rhythmic movements are particularly well suited to coordinate behavior and create a cognitive space that combines presence, coordination, and meaning.[30] These forms of collective performance could then gradually turn participants into audience members insofar as they develop a sense that they are performing a ritualized and thus repeated action.

Narrative Elements of the Early Stage

On the basis of joint attention, performances, and receptivity, narrative communication can take place. In the words of a previous chapter, we need to learn how to track characters, experience them as playable from within, and explain what they are doing. Collective performances on stages may have trained and shaped the way we engage in narrative thinking in several specific ways.

Beginning and ending. In order for a performance to occur, there must be signals to the audience that things are about to begin. These can be signals given by the performers and narrators to attract attention. In the formal theater setting, this is usually done by music, the curtain, and the lights. The audience also has its part in setting the marks of beginning and end: they become quiet and turn their attention to the stage. In general, their coming together, the assembly, is already an essential demonstration that there can be joint attention. And it's the same at the end. The release of the collective tension is evident in the movements of those gathered. The group breaks up. They disperse. Similar signals of beginnings and endings occur in storytelling, too. "Once upon a time . . ." sets clear expectations. We lean back. Someone—the narrator—has a story to tell. We pause and wait. Everyone stares at the storyteller and expects something.

The transformation of real people into characters. One of my earliest theater experiences was a performance by Marcel Marceau, the mime. I still remember exactly how he could change his personality within fractions of a second. For example, he would disappear behind a small wall, only to emerge immediately on the other side in a different role. As Goliath, he could chase David, while only running in circles, and in a flash he was a different character as soon as he emerged from behind the wall again. Each of his gestures corresponded to the new role. Even if I had had no conception of the theater at that time as a child, I would have noticed that he, the same actor, appeared as someone else. I could at least sense that there was a difference between his person with its internal sensitivities and the state he showed to the outside world. The perception of this difference is the first step in understanding casting—that is, the way someone slips into a role.

Following Martin Heidegger's suggestion that a basic form of thinking is that we can think "something as something," one could say that stages help us learn to interpret *something as something (else)*, and likewise *someone as someone (else)* and *an action as (another) action*.[31] We know that there is someone in front of us, who now acts "as" Goliath.

Such a demonstration of something as something (else) might be particularly striking if a human being imitates an animal. If a man crawled on all fours, even though his clan knows quite well that he has healthy legs, it could perhaps occur to them that he was imitating an animal. The clan might detach and decouple the represented from the person representing it.[32]

The same applies, of course, to masquerades with fur and feathers or face painting. Playing animals might have particularly promoted the detachment from the present and encouraged imagination, because animals display such stark differences from humans. In this respect, animals are more playable than humans. Perhaps this is also a reason (besides the importance of hunting) why animals are particularly represented in the depictions of cave paintings or in early sculptures.

The distinctiveness of animals promotes mobility of consciousness. Whether the playability of animals was also associated with religious significance is a question we do not need to pursue here.

Transition from the here to the there. When performers play a character such as an animal, the actors are simultaneously physically present and still represent something else. This balancing act between reality and the imaginary is probably impossible for non-human animals. When I crawl on all fours in front of my cats, they do not see my attempt to play a cat. It follows that they may not be able to exchange concrete past experiences with each other, at least not by reenacting a past episode or by symbolic communication. Narrative thinking, on the other hand, motivates our mobile consciousness to leap into absent, imaginary worlds.[33]

When a performer on a stage flinches, this jumps to the audience and becomes its flinch. But when the performer flinches from something that is not really threatening—for example, another performer portraying a lion—it's as if the audience is flinching from the imaginary world. The audience is in both its own physical world and this world of the imaginary. This effect is, after all, precisely what makes theater so appealing—and the physical presence of many spectators can reinforce the impression of the performance when emotional reactions resonate and are amplified by many.

Repeatability. When a performance is over, the sequence as a whole is mentally marked as a complete episode that has led to a specific end. This episode can now be remembered and thus recalled. Students can repeat the act their mentors taught them. In the theater, a play can be performed again by the same or different actors. Audience members can review the play in their minds. They can also mentally try out a different version: "What would have happened if . . ." Stages prepare the actual and mental repeatability by ritualizing the sequences of actions. To extend this idea, we could suggest that the mental simulation that we produce of our real situation is a move to make our world a stage that we can rehearse and repeat in our mind. Repeatability also opens the door for playability and for change.

Emotional response. It's one thing to watch a movie alone, and another to watch a movie in a group. Horror flicks, for example, almost require the collective sensorium that stimulates reactions ranging from giggles to shrieks. Likewise, storytelling in front of others can captivate everyone. When an exciting or sudden event occurs, the audience may gasp. When the gasp resounds through the rows of spectators, the emotional reaction is amplified. Even someone who was not paying attention will get pulled in.

An actor on the stage makes a frightened leap and the alarm can spread to the audience. The actor jumps back in front of an imaginary leopard and the audience gasps at the sudden movement, even if it seems to happen without reason. Suddenly, the audience members feel the presence of the leopard because their shock shows them retroactively that there was something that triggered the jump. A leopard emerges in their mind, and they have entered an imaginary situation. The emotion of the shock operates as the hinge connecting the worlds.

Numerous emotions are contagious. In addition to horror, we also have laughter. When one person starts laughing, others usually laugh along and thereby intensify the effect of the performance. Emotions that are shared with others are particularly strong and memorable. Typically, a narrative episode ends with an emotion. From the perspective of the shared performance on a stage, emotions gain an additional positive dimension in that they are shared by everyone and therefore unite the group. Many of the rewarding emotions discussed earlier were likely connected with collective experiences in the first place. Laughter as a defusing of embarrassment needs an audience. Satisfaction in moral misdeeds seeks recognition from others. We revel in triumph as a public display.

The site of the stage. The stage contours the zone in which the performers move. This space possesses a certain aura and the objects on this stage are special: the objects stand for something else in this space, this place of elevated performance. This is true for spontaneous stages as well as institutionalized stages. Sacred spaces are particularly marked by magic or aura because shamans or priests perform something that aims at invisible forces; kings and queens on their thrones evoke the history of the state; and even academics in their lecture halls can occasionally mesmerize crowds with their stories. In the Christian

Eucharist a past history is remembered and embodied at the same time through bread and wine. In spontaneous "stagings" all found objects can suddenly become props of a performance and acquire a new meaning. Everything is possible for the storyteller; we are where multiversionality is in the air.

What Have We Learned?

To understand how immersion or transportation into story worlds is possible, we looked more closely at the concept of "mobility of consciousness" to describe how humans mentally co-experience what others perform for us or tell us in the form of stories. The thesis of this chapter was that "stages" might have played an important role in practicing narrative thinking and the mobility of consciousness.

Our task here has not been to sketch a cultural history of the theater as an institution. Instead, we have identified some basic elements that may have been part of the development of stages and may have facilitated the learning of narrative thinking. That these considerations are hypothetical does not need to be emphasized here. The goal was to isolate the cognitive and cultural elements that might have supported the development of the mobility of consciousness and narrative empathy. This brings us back to the central question of how we humans have acquired the ability to detach our consciousness from our bodies and to empathize with others and their situations by means of narratives.

Here are the inferences that follow from our inquiry:

1. The co-experiencing of the situation of others did not arise "by itself" on the basis of observation of everyday situations. Rather, a range of cultural techniques had to be developed, such as staging and performing events for others.
2. Such targeted performances create the ritual space of the "stage." It is a ritual insofar as the attention and behaviors of the participants are coordinated and repeatable. The participants have the feeling that something relevant to them is happening here, something that demands their attention.

3. Participants include both actors and audience. Both sides have expectations of each other about their roles and the performance. They form a powerful dyad (actor-audience, performer-spectator, speaker-listener, and the like).

4. The collectivity of the spectators contributes significantly to the success of the transportation. Spectators infect one another with their emotions. Even astonished silence has to be learned. Collectivity gives security, marks the beginning and end of the performance, and underscores the emotions of the performance.

5. Similarly, the rhythm of musical performances or dance helps to synchronize actors and audience. All can become one in collective dance.

6. Mimetic performance can help facilitate the transportation for the audience from the physical presence to the imaginary world of the story. The representation of animals, for example, emphasizes the strange behavior of the performer and underlines the "something as something (else)" of the performance.

7. In collective performances, the performers can serve as a medium for the spectators. When the hunters being portrayed are frightened, the audience is frightened as well. The performance produces an effect of presence that draws the audience into the present moment of the staged story. In narratological terms: discourse and plot merge. The spark of the performance on stage opens the gate to an imagined scenario beyond the stage.

8. In other words, mobility of consciousness is made possible when other people (or characters) make themselves available as carriers of consciousness. This is especially so when the audience members who are to empathize do not act themselves but are only observers and spectators.

9. Without such stages and ritual performances, we would probably have trained our mobility of consciousness less successfully.

10. Speculatively, we have considered the hypothesis that our consciousness *is* essentially a mobile one, which means by extension that it is, by definition, open to narrative and empathy. (This does not determine whether a different form of consciousness without mobility—and thus also a non-empathetic consciousness—is

possible. If such a consciousness without mobility were to exist in another form of life or in a machine, it would be very strange and alien to us.)

11. Narrative thinking owes its existence to a *decoupling* from the physical presence to the space of the imagined. Decoupling releases the creative processes we associate with narrative thinking, the playing of characters, the multiversionality of events, and the emotional rewards that narrative thinking promises. The stage allows the leap to this disconnection.

12. Taken together, mobility of consciousness, stage performances, co-experience, and narrative thinking are likely to have formed and developed reciprocally and in sync with each other. The stage was the hearth that kindled our collective imagination and experience.

EPILOGUE

As humans, we are narrative beings. Our stories change how we experience our lives. Although we cannot control all events in the real world, we are in charge of our own stories, and we can rebel against stories imposed on us. Narrative thinking allows us to transform the way we experience, remember, and share our stories. Our engagement with stories does not feel like an effort. Rather, we love to engage in narrative thinking, as it promises us stories that end with an emotional reward. In this book, we have encountered a wide array of such emotional rewards, ranging from the classic happy ending to the therapeutic resolution of collective crises, and from joyful surprise to the satisfaction of punishing villains. It's all about emotions. However, what's important here is not simply the emotion as a fuzzy feeling. Rather, the emotional ending of a narrative episode signals to everyone that the narrative is over for now and that we can mentally "return" to ourselves and our present situation.

When we engage with narrative thinking, we are transported into a different world, a story world. This occurs in grand fiction and movies, but it is also true for everyday chitchat and the micronarratives we come across in our fantasies. In this book, we called our human capacity to "move" into these narratives "mobility of consciousness." Good narratives hijack us; we're placed in new situations and play a character. And that means we see ourselves right in the middle of the story. Once a story is over, we may look back on it, but the exciting part where we co-experience another's situation is when everything is still possible: we do not yet know what will come to pass. Even when we remember an episode from our past or watch a movie for a second time, the excitement lies in what is or was once possible. Narrative thinking makes us creative. We imagine and anticipate various scenarios of things that might occur. We align our fears, hopes, wishes, and guesses with the hard facts of the story. This means narrative thinking or story thinking always comes in the plural. There's not just the story of what happened, but the "multiversionality" of what could happen.

The explanations and the evidence for these ideas can be found in the chapters above. There are two aspects left for me to discuss here at the end: I would like to acknowledge the good work of narratologists who have chosen different starting points than I have, and I want to make known my fear that we may be losing our narrative thinking in the present state of our culture and technology.

The Two Camps of Narratology

Narratologists tend to fall into one of two competing camps. On one hand, there are theories that center on the notion of the event. On the other, there are theories that take their start from Gérard Genette's notion of the voice (*voix*) and choose narrative perspectives as the basis for the analysis of narration. Each of these camps is characterized not so much by a method as by this focus. Computational narratologists and philosophers, for example, tend to focus on events, and literary scholars and cognitive narratologists on perspectives, but they can be found on both sides of the divide. Both have created excellent insights.

First, let's take a closer look at the event-focused theories of narratology. In these, the radical transformation of a represented state is central.[1] According to these theories, a narrative only occurs when an incisive event exists or is mentally contemplated that fundamentally changes or could change a situation. Wolf Schmid, for instance, argues that four conditions must be met for such a narrative event: (1) relevance, (2) unpredictability, (3) persistence, and (4) irreversibility. Let's choose a small episode to illustrate this:

> A girl walks home from school. There is a chestnut tree on the street. She picks up a beautiful chestnut and throws it over a fence. There is a loud clatter and the sound of a window shattering. The girl is frightened and flees. She arrives home with a pounding heart. No one has seen her throw the chestnut, but she feels different from now on. The next day she doesn't leave the house and refuses to go to school.

EPILOGUE

For the girl, this event is relevant (1). Even though she may not have to fear any consequences, she did not expect it (2). The event continues because she feels guilty (3), and she considers the event irreversible (4). From here, even without being a Poe, Balzac, or Woolf, one can construct a story in which this very small event leads the girl to see herself as *defined* by this event. Every event could thus become a fundamental turning point, a trauma, or an edge, and that is precisely what events are.[2]

If you're looking for a sober summary of this definition, you could say that narratives are strings of events, similar to how Vladimir Propp and many folklorists have approached narratives. Events bring about important, lasting, and irreversible changes that were not clearly recognizable in advance. Thus, they could not have been predicted from calculable natural laws, even if such a law of nature might be found in retrospect, such as a snowmelt leading to an avalanche, but events still have patterns that we can describe and classify.

Events are decisive and may be traumatic or liberating. One could also emphasize the paradoxical aspect of the definition and say that the representations *before* the event and *after* the event are incompatible and mutually exclusive. The event is a caesura.[3] Through the event, two states are held together, such as two ways of seeing the world, two different moods or scopes of action that cannot be united. The girl before throwing the chestnut and the disturbed girl afterward are different beings. When someone wins the lottery, his outlook on life and the world is fundamentally changed—or it was not an event after all. When someone makes a new friend, that can be a true change of life (and therefore an event) too, even if we are not able to clearly name all the points of the change. Terrible accidents or illnesses plunge a person and witnessing observers into a different reality in which previously pressing concerns become unimportant. Only in retrospect can both sides of an event, the before and after, be united. However, at that point, we have left narrative thinking behind and have moved on to something else. The episode gets filed away; we have moved to the next segment. This is an important aspect of narrative thinking—that it can cease, become inactive. In itself, however, narrative thinking has an unfinished nature.

Walter Benjamin drew a central consequence from such narrative events: they are the rough surface that interpretation aims to grab. An event does not have a specific interpretation, but it can be interpreted, reinterpreted, and framed in many different ways.[4] Interpretation is the attempt to reintegrate the event into the flow of a clear order, thus ironing out and nullifying the event. The caesura of the event is to become a bridge that unites the before and after. Narratives have an index toward coherence—they motivate the search for coherence, but without already presenting it. Nevertheless, interpretation does not come to the narration "from the outside." Rather, the event only becomes an event because an attempt is made to integrate it, which already implies some interpretative work.[5]

From here we can see how narrative thinking would appear if we were determined by event narratology and constantly on the hunt for events. A question would be what we would do when no major event is available. Two things: we would first turn on Netflix, and then we would make mountains from molehills. As is well known, for a man with a hammer, everything looks like a nail. Accordingly, for our narrative thinking every little incident would be an occasion for a radical turnaround. Our mental casting scouts would always be on the lookout for suitable actors in our next event-focused drama: every boss would seem like a tyrant to us. Like an Othello, we would suspect every lover and friend of betraying us. Moreover, we would constantly be searching for the (next) great love. Perhaps this describes us well. As humans, we are rarely satisfied with calmly grooming each other in the savanna, but prefer to act like drama kings and queens.

Now let's take a look at the other camp of narratologists who start from the voice, discourse, and perspectives on events and facts. Besides the term *perspective*, the term *representation* is regularly used. We could say that this narratology (or collection of narratologies) is concerned with the representation of facts. The basic experience here is that the same facts presented from different perspectives become different facts. The mood of the narrator, to give an example, frames what has happened in such a way that the whole account takes on a different color and thus creates a different image. Gérard Genette has developed this insight into a systematic classification system that describes the most diverse narrative perspectives.[6] The apparent static nature of the classification system should not obscure the fact that the most diverse forms of representation fundamentally shape a narrative, such as "free indirect discourse" or different forms of the implied audience. The various speech positions, forms of representation, and perspectives cannot be translated into each other.

To illustrate this point, I will use an anecdote from my lab. In one of our studies (not yet published), we wanted the participants in our experiment to transfer lessons from stories to everyday-life situations. Specifically, we presented participants with Aesop fables and short scenarios of modern life dilemmas, such as the question of whether you should quit a job if you have a difficult boss. Some of our fables had explicit advice in this regard (in this case, some fables warned that problems exist elsewhere too, and staying might be wiser; other fables suggested that a bad boss will always be bad like a wolf, and it's better to leave sooner rather than later). Participants usually could transfer the lessons from these stories well, and they also found the fables highly persuasive (they stated that they would do the same thing, too, regardless of what the advice was). However, we had some stubborn fables that would not submit to the will of our experimenting minds. In a variety of fables, we did not succeed in reliably getting the participants to transfer the learning lesson correctly to a life situation. On closer examination it turned out we had counted on a change of perspective that the participants did not want to go along with. In the fable "The Fox and the Raven," for example, our participants could immediately understand: Don't fall for flattery. But despite all our attempts, the participants were mostly unwilling to turn this lesson around: If you want to fool someone, use flattery. That would seem to be a different story for them. The perspective of the story at the end (the victim raven) is a central part of the story. What this anecdote shows is that we witness a story from the inside, from one specific perspective, and resist considering another. In this respect, the perspective and also the voice from which and with which the story is told are not detachable aspects of a narrative, but its basic foundation.[7]

I will add a more explosive reflection on the importance of narrative perspective. When George Floyd died in Minnesota in 2020 as a result of police violence that was later ruled to be murder, the video recording went viral around the world. Black Lives Matter was suddenly on everyone's lips. What interests me in this context is this: I wonder from what perspective people perceived this video of George Floyd's choking death—from that of Floyd or that of a sympathetic observer? From the former perspective, observers would slowly co-experience the suffocation. From the latter, they'd wish they had been there and could have intervened.

I would bet that the perspective makes a difference in how people will keep this event in mind in the long run and how much it leads to a change in actual behavior. Observing the murder from the internal perspective of the victim likely evokes despair; co-experiencing such a death is simply horrifying. In contrast, taking the perspective of an observer calls up more positive tendencies to encourage intervening actions. People may wish they could have done something. This seems positive, but at the same time there is the latent danger that one's own mental commitment will turn into some form of deceptive self-praise, since one might feel one deserves praise for simply wanting to do the right thing. I have characterized this form of helper-identification as "false empathy" since its effective result can be to feel good about oneself: standing on the right side and opposing police violence is good, and the victim is, to overstate this point, a regrettable but acceptable sacrifice to bring about this effect.[8]

Let's return to perspective-focused narratology. The argument for this narratology, which I have constructed here via these case histories, is that narration from one perspective is cognitively different from narration of the same plot from another perspective. For this narratology, the differences between human beings, theory of mind, is the horizon of our thinking.[9]

Although this book's approach to narrative thinking diverges from those that center on the "event" or "voice," I do not wish to dismiss the excellent work of these scholars, narratologists, and scientists. The readers of this book will get to choose their starting point to make sense of our human capacity.

Should We Turn Against Narrative Thinking?

Many scholars caution against narrative thinking. It is associated, for example, with fake news, polarization, and extremism. In conflicts like the wars between Ukraine and Russia or Israel and Gaza or in tense standoffs like those between India and Pakistan or China and Taiwan, competing narratives declare who is to blame. If we follow the Russian narrative, for example, the country has suffered humiliations and loss of its unity. Ukraine is a limb of Mother Russia, severed off by the West, that needs to be reunited with the homeland. Many narrators can tell their own tale as a story of vulnerability. There are victims on all sides, and focusing on victimhood and vulnerability works well to establish righteousness for everyone. The consequences are horrific.

Accordingly, Jonathan Gottschall defines stories as persuasion machines. Stories create questionable notions of war and peace, gender images, and the future that take root in our minds, generate harmful stereotypes, and guide our behavior. Fake news spreads faster than real news. Narratives, we can suspect, determine our lives precisely because we do not notice how we are controlled by them. The influence of stories on our lives, seen in this way, can be disastrous. Gottschall begins his book with the multiple shootings at a Pittsburgh synagogue in 2018, where, as Gottschall reports, the assassin was driven by age-old antisemitic narratives.[10] In recent years we have borne an exponential growth in the list of victims of antisemitism. Even Plato warned against fictions and narratives, famously considering banishing poets from his republic for spreading false ideas.

So, should we banish stories, limit narrative thinking, and guard ourselves against it? I don't think so. On the contrary, I think we need more narrative thinking and need to learn to develop our story thinking. We need more, not fewer narratives and more, not fewer perspectives. The dangers lie in a poverty that would result from reduced narrative thinking.

In the case of the ongoing wars, for instance, I see a grave danger in the reduction of narrative perspectives. In Year 1 of Russia's war against Ukraine, the Western media still offered many narrative perspectives, including those of the blue-eyed Russian soldiers who did not know what war they were tossed into. In Year 2, the entire conflict tended to be reduced to just a few perspectives of the figureheads—namely, Putin (evil and determined) and Zelensky (exhausted, but still heroic). In Year 3, Putin is as evil as ever, while Zelensky is aiming to reclaim agency. When there are few perspectives, possibilities for peace slip away, since no one seems to be thinking of a compromise between two polarized perspectives.

With fanatics on all sides, single narratives block out all others. The narratives told by many Israeli settlers (as in the powerful documentary *The Settlers*, directed by Shimon Dotan, in 2016) and the narratives told by members of Hamas only overlap in that they both concern the same soil. These single narratives are not a basis for compromise or co-existence. Yet every narrative contains possibilities for plurality, for other perspectives, and multiversional outcomes. Hope and an open future exist not despite but because of narrative thinking.

My concern is that we might unlearn narrative thinking, that we might fall into a kind of narrative deprivation, a weakening of our ability to experience the world in stories. This narrative poverty may well be accompanied by narrative naiveté, a loss of the sense that any story has the makings for new stories, the potential for alternative versions.

At first glance, this concern seems paradoxical, since we live in a world of stories and spend many hours a day in and with stories.[11] Stories are extremely well suited to give thought processes a form that links our current state with expectations for the future. I don't doubt that stories are ubiquitous. Every advertisement makes us a narrative offer. But I worry that too many of the stories in our lives are not created or mentally co-created by us. Artificial intelligence like ChatGPT will get better and better at telling our stories for us. The abundance of stories on social media and the perfection of the stories all around us can make us lazy. Schooling does not emphasize story thinking. Narrative thinking is not about running through a given story, but about thinking further, discovering what's possible within it, and experiencing multiple versions of it. It's about the exit from self-incurred narrative immaturity (to hint at Immanuel Kant). It's precisely here that my concern has its origin.

It has never been easier to immerse oneself in one-sided narrative worlds at the touch of a button. New technologies have brought narrative to life via AI, video games, and virtual reality (to be sure, I am mostly positive about these technologies).[12] The speed of storytelling is also significant. Ernest Hemingway could, allegedly, win a bet by telling a story in six words.[13] TikTok presents a vast number of ministories that manage with six seconds. (Harmut Rosa speaks of social acceleration as the core characteristic of our epoch.)[14] This overabundance of stories, however, does not constitute narrative thinking. Narrative thinking tries to bring an episode to an end *and* at the same time to engage in multiversional thinking. For every story there is an alternative story, so there is no *one* narrative. Opening up to plurality can certainly succeed in new media and on platforms like TikTok and YouTube, because there are many alternative stories for every story, if you can find them, and everyone can add his or her own version. Every story has more than one perspective. However, this ability also needs to be encouraged and trained. Anyone who is manipulated by a story has not thought it through to the end and has not witnessed the alternative versions. The synagogue shooter obviously could only think of his side of the story, not the versions of horror and suffering that go with it. He simply took the antisemitic narrative as truth. This can perhaps be described as manipulation, but not as narrative thinking.

In this book, I have suggested that narrative thinking is dynamic thinking that balances two tendencies. On one hand, there is thinking in episodes that end with an emotion that rewards the audience. Episodes are complete when they connect a beginning and an end. Such episodes provide stability and orientation in a complex world. On the other hand, there is thinking in terms of possibility: in every story, everything could turn out differently. Stories are experienced from their middle and not from the end. The question is how it could end, not how it did end, because each ending is only one possible outcome of many. This thinking is multiversional because it continually generates probable or less probable, hoped-for or feared versions of what might happen or has already happened. Narrative thinking thus sensitizes us to contingency: everything could always turn out differently.

Are there alternatives to the narrative brain? What kind of thinking would not be narrative? This book has alluded to some alternatives to narrative thinking, including rational-causal thinking, thinking in fixed identities, thinking in images, the flow of unstructured daydreaming, and the frenzy of permanent presence. All of these alternatives are similar in that they can emerge from narrative thinking or, conversely, can lead to it. Each of these forms of thinking has its charms, but it's difficult to identify an actual alternative. Thinking in terms of causality is explanatory, but not fulfilling, since it lacks the emotional structure that gives meaning. Thinking in terms of identities and fixed images threatens to lead to pathologies and fixations. Daydreaming and the frenzy of permanent presence lack structure that makes them bearable in the long run. Some readers may see it differently. For me, however, hope lies in narrative thinking.

Those who engage in narrative thinking find themselves in the tension between the two tendencies described above: recipients want to bring the story to its respective end and reward themselves emotionally in the process, and yet they continually recognize the multiversional forks in the road at which the stories could develop in other directions. The result of this tension is intensity. Intensity, to forestall all misunderstanding, consists not in a fulfilled moment or a dull presence, but in a tension that could be discharged in many directions. To take up the image of intensity conveyed by etymology as the tension of a bow: intensity rests in the bow before it discharges.

This intensity has many dimensions: it binds us to life and gives it meaning. It increases the sense of presence. The aesthetic experience is intensified, for what is aesthetics in narratives but an intensification of the moment, since this moment is seen from many perspectives and understood as the junction of many developments? Intensity sharpens our attention and gives weight to the turning points at which decisions are made. The morally better choices are strengthened because the moments of narrative intensity are relived again and again— and who wants to have to relive wrong choices over and over again? Intensity prepares for many situations and then allows for the quicker or better response. Narrative intensity also has a distinctly dialogic dimension because it involves different perspectives. It introduces one to the lives and views of other people whose stories are being witnessed. People have empathy because they hope along with others in a state of uncertainty. Understanding other people, theory of mind, thus also belongs to the field of this narrative intensity: understanding other people does not simply mean guessing their concrete intentions and feelings at a moment, but rather understanding which possibilities, uncertainties, and decisions resonate in their actions and feelings—that is, in which narratives they see themselves entangled. In other words, without narrative thinking and the intensity it feeds, empathy becomes shallow.

The list of benefits derived from narrative thinking is long. I'm not cut out to be a prophet, or I would have started my own cult by now, but here's some good news: narrative thinking can make you happy. Those who live in a world of gossip usually fare better.[15] Even nasty gossip is only really dangerous when people are systematically excluded and its narrative dimensions are being stymied. Those who can recount their everyday life in narrative episodes can process it better. For a narrative thinker, even negative experiences can lead to gain. People are unhappy in a world without narratives and stories—that is, without the knowledge that they live in stories—because it's a world without alternatives, without sense and meaning and without depth. It's also boring.

This book arrives at a simple conclusion: We should encourage narrative thinking, because it does not thrive on its own and can sometimes veer into strange or dangerous directions. The best way to nurture it is to tell stories ourselves, to give narrative form to everyday life and to the great questions of our lives. Every story contains multiple potential versions, each opening new possibilities, and these come alive through the act of retelling. This is an exciting experience for children in school as well as for adults: to realize how one's own seemingly truthful retelling changes an event. Even the hopeless situation, the misfortune, and the long-believed collective narrative can be bent around as soon as they appear in the form of one's own narrative and become playable. Of course, this does not mean that one should provide a better ending to every story à la Hollywood, or that truth and facts do not matter and one makes the world as one pleases.

But it does mean seeing spaces of possibility and witnessing the perspectives of others. As *Homo narrans*, we aren't just beings who are being told stories, but beings who tell and retell stories ourselves. All storytelling takes place in the plural. A narratively enlightened consciousness is one that is willingly captivated and guided by narratives but at the same time is aware of the unleashing potential of stories. In every story there is the escape and exit from a world that is perceived as too narrow. To make this storied getaway, I pin my hopes on narrative flexibility.

ACKNOWLEDGMENTS

This book has presented a number of emotions that reward us for our narrative thinking, and thus each represents the end point of an episode. I have omitted one important emotion: gratitude. The reason for this omission is that gratitude is different from the other narrative emotions. On one hand, gratitude also stands at the end of an episode, but on the other, it also marks a future and a new beginning for continuing to play with changed roles. Gratitude is an ending without an end.

Many have helped me in the writing of this book and my heartfelt gratitude goes to them. My hope is that we will continue the conversations and that they will continue to be an impetus of happy stories: Colin Allen, Shahzeen Attari, Frauke Berndt, Johan Bollen, Urs Breitenstein, Kira Breithaupt, Leela Breithaupt, Rüdiger Campe, Michel Chaouli, Yulia Chentsova-Dutton, Anna Chinni, Bettina Christner, Wolfram Eilenberger, Eva Esslinger, Samuel Evola, Nicola Gess, Eva Gilmer, Robert Goldstone, Manuel J. Hartung, Milo Hicks, Benjamin Hiskes, Douglas Hofstadter, Philipp Hölzing, Christoph Irmscher, Günther Jikeli, Friedemann Karig, Suzanne Keen, Cameron Kincaid, Charlotte Klonk, Sara Konrath, Albrecht Koschorke, Teresa Kovacs, John Kruschke, Sandra Kübler, Victoria Lagrange, Binyan Li, Winfried Menninghaus, Jillian Meyer, Ege Otenen, Eyal Peretz, Benjamin Robinson, Hartmut Rosa, Elizabeth Schechter, Eleanor Schille-Hudson, Thomas Schoenemann, Richard Shiffrin, Jan Standke, Jan-Erik Strasser, Yiyan Tan, Philipp Theisohn, Dieter Thomä, Peter Todd, Elvira Topalovic, Johannes Türk, Christian Weber, Philipp Weber, Marc A. Weiner, Arne Willée, Claire Woodward, Devin Wright, Rüdiger Zill, Lisa Zunshine, and all members of the Experimental Humanities Lab.

Brian Donarski is the secret co-author of this book and my copyeditor. Wolfram Eilenberger is my adviser in all last things.

The English version of this book, substantially altered from the German edition, benefited greatly from the tireless encouragement of Jean Black and Elizabeth Sylvia at Yale University Press. Elizabeth Casey gave the text its final form. There are many crisp and fortuitous formulations that found their origin in her pen. The revisions have been made possible by a fellowship at The New Institute in Hamburg, Germany, for 2023–2024. Several of the cited studies and my Experimental Humanities Lab have received funding for participants by Indiana University's College Arts and Humanities Institute (CAHI) and the Indiana University's Cognitive Science Program.

NOTES

Introduction

1 Schilbach, "On the relationship of online and offline social cognition."
2 See also Kelley et al., *An Atlas of Interpersonal Situations*. The authors go one step further than the definition provided above by already explicitly including outcomes in the definition of situation. In my definition, the uncertain outcome is part of the concept.
3 For an introduction to neuroscientific measurements, see Hohwy, *The Predictive Mind*. On predictions, see Hutchinson and Barrett, "The power of predictions."
4 See Clark, *The Experience Machine*.
5 Zanger, *Film Remakes as Ritual and Disguise;* see also Breithaupt, *Kultur der Ausrede*.
6 Zacks et al., "Perceiving, remembering, and communicating structure in events."
7 Wegner and Gray, *The Mind Club*.
8 Stawarczyk et al., "Event representations and predictive processing"; Sargent et al., "Event segmentation ability uniquely predicts event memory."
9 Green et al., "Understanding media enjoyment"; Keen, "A theory of narrative empathy."
10 See Mesoudi, *Cultural Evolution*.
11 Dawkins, *The Selfish Gene*.
12 For an overview, see Laland and Sterelny, "Perspective."
13 Fuchs, *Ecology of the Brain;* Bobbice, *History of Emotions;* Roskies, "How does neuroscience affect our conception of volition?"
14 There are, to be sure, possibilities in this domain. One such attempt has been made by Paul Armstrong, who chooses primarily the temporal inequalities of neural processes as the starting point for his ideas; see Armstrong, *Stories and the Brain*.
15 Glasser et al., "A multi-modal parcellation of human cerebral cortex."
16 Wolf, *Proust and the Squid*.
17 Bartlett, *Remembering*.

Chapter 1. Thinking in Episodes

1 Musil, *The Man Without Qualities*, 3.

2 Bornstein et al., "Perception of symmetry in infancy."

3 Famous and still touching is the story of how Konrad Lorenz raised the goose child Martina, who imprinted on him; see Lorenz, *Er redete mit dem Vieh, den Vögeln und den Fischen*. Wood and Wood, "One-shot object parsing in newborn chicks."

4 See Luhmann, *Introduction to Systems Theory*.

5 Hendriks-Jansen, *Catching Ourselves in the Act*, 306–309.

6 Tulving, "Episodic and semantic memory."

7 Tulving, "Episodic memory: From mind to brain."

8 For an overview, see Crystal, "Elements of episodic-like memory in animal models."

9 The distinction between animate and non-animate is very basic and recognized by children at an early age; see Opfer and Gelman, "Development of the animate-inanimate distinction."

10 Another aspect is Benjamin Libet's observation of the temporal delay of consciousness; see Libet, *Mind Time*.

11 Sargent et al., "Event segmentation ability uniquely predicts event memory."

12 Magliano and Zacks, "The impact of continuity editing in narrative film on event segmentation"; Pettijohn and Radvansky, "Narrative event boundaries, reading times, and expectation."

13 Brigard, "Is memory for remembering?"; Dennett, *Consciousness Explained*; see also Zacks et al., "Event perception."

14 In contrast stands our so-called default network—the state that the brain sinks into without much energy expenditure; see Raichle et al., "A default mode of brain function."

15 Radvansky et al., "Walking through doorways causes forgetting"; Radvansky and Zacks, "Event boundaries in memory and cognition"; Radvansky and Zacks, *Event Cognition*.

16 Radvansky and Zacks, *Event Cognition*.

17 Franklin et al., "Structured event memory."

18 I thank Anna Chinni for the suggestion of *The Giving Tree*.

19 Monroy et al., "Translating visual information into action predictions."

20 Speer et al., "Human brain activity time-locked to narrative event boundaries."

21 Chang et al., "Relating the past with the present"; Chang et al., "Information flow across the cortical timescale hierarchy during narrative con-

struction"; Dominey, "Narrative event segmentation in the cortical reservoir"; Hahamy, "The human brain reactivates context-specific past information at event boundaries of naturalistic experiences."

22 Radvansky and Zacks, for example, argue that people track the causal structure within an event and use it for memory; see Radvansky and Zacks, "Event boundaries in memory and cognition." Intuitively, causality is important, but even so, the evidence is somewhat sparse. More on this later.

23 Nabi and Green, "The role of a narrative's emotional flow in promoting persuasive outcomes."

24 This was Frank Kermode's big question; see Kermode, *The Sense of an Ending*.

25 Aristotle, *Poetics*, Book XI.

26 Freytag has since fallen into disrepute because of his use of antisemitic stereotypes in his novel *Debit and Credit* (1855).

27 Freytag, *Die Technik des Dramas*, 105–190.

28 Freytag, *Die Technik des Dramas*, 97–105. Freytag understands his thesis here as a transhistorical description of both ancient and modern Western drama. Obviously, we can object here that concepts of individuality have changed and the idea of an inner life of a character has gained traction in Western modernity; see for instance Bloom, *Shakespeare: The Invention of the Human*.

29 Relating to the "outflow and inflow of willpower," "proposition and opposition," or "the becoming of the deed and its reflections on the soul," see Freytag, *Die Technik des Dramas*, 97.

30 Freytag, *Die Technik des Dramas*, 98, 99, 100, 98.

31 Frijda, "The psychologists' point of view," 69.

32 The writer E. M. Forster sparked a long discussion in narratology about whether a sentence like "The King died and then the Queen died" was already a narrative; see Forster, *Aspects of the Novel*. Forster decided against it, only to add that the sentence becomes a narrative if we make an addition: "The King died and then the Queen died out of grief." This addition would be more in line with Freytag's requirement, since the queen's grief implies an emotional reaction and thus an inner life. Some objections were raised against this, such as that the first sentence also constitutes a narration, insofar as the recipient can create mental links.

33 Much has been written on this paradox of the positivity of negative emotions; see, for example, Hanich et al., "Why we like to watch sad films."

Chapter 2. Telephone Games

1 Kashima et al., "The maintenance of cultural stereotypes in the conversational retelling of narratives."

2 Dudukovic et al., "Telling a story or telling it straight."

3 Tobin, *Elements of Surprise.*

4 Varnum and Grossmann, "Cultural change."

5 Stereotypes are also reinforced; see Kashima, "Maintaining cultural stereotypes in the serial reproduction of narratives."

6 On schema formation and abbreviation, see also Mandler and Johnson, "Remembrance of things parsed"; Koschorke, *Fact and Fiction*, chapter 1.

7 The emphasis on social information in narrative transmission is already well established; see Mesoudi et al., "A bias for social information in human cultural transmission"; and Stubbersfield, "Serial killers, spiders and cybersex."

8 Bartlett, *Remembering*, v.

9 See Mesoudi and Whiten, "The multiple roles of cultural transmission experiments in understanding human cultural evolution." Research in recent decades has found that bizarre elements survive more clearly in later generations of narration (Nyhof and Barrett, "Spreading nonnatural concepts") as do strong emotions (Heath et al., "Emotional selection in memes"; Stubbersfield et al., "Chicken tumors and a fishy revenge") while information generally declines (Griffiths et al., "The effects of cultural transmission are modulated by the amount of information transmitted").

10 For reinforced transmission of schemas, see Rumelhart, "Understanding and summarizing brief stories."

11 Bartlett, *Remembering*, 85, 86, 84.

12 See Reber, *Critical Feeling*, on "ease of processing"; Bergman and Roediger, "Can Bartlett's repeated reproduction experiments be replicated?"; Mandler and Johnson, "Remembrance of things parsed"; and Rumelhart, "Understanding and summarizing brief stories."

13 On the reception and neglected aspects of Bartlett, see Wagoner, *The Constructive Mind*; Mandler and Johnson, "Remembrance of things parsed."

14 Varnum and Grossmann, "Cultural change."

15 How stories are passed on and in what sense they are collective can vary significantly. During a field study in central Australia, I was told that the myths of many indigenous inhabitants of Australia are regarded as secrets. Those who receive a story are charged with keeping that secret and passing it on only before death, thereby passing on responsibility for the

secret. Thus, each individual knows only a few details of the overall story. These stories are collective in a different sense than we are used to in the West. No individual knows the collective and common story, but as a group they do. I am certainly not an expert in this case, but I was intrigued by the idea that stories could operate not as a common good but as secrets, with each member of a community being a secret keeper.

16 Grätz, *Das Märchen in der deutschen Aufklärung*. In addition to Grätz, other particularly illuminating accounts include Zipes, *Fairy Tales and the Art of Subversion*; Tatar, *The Hard Facts of the Grimms' Fairy Tales*; and Norberg, *The Brothers Grimm and the Making of German Nationalism*. It is also important to consider the concept of non-excellent, mediocre art, or indeed the non-art of the time; see Fleming, *Exemplarity and Mediocrity*.

17 Sheehan, *The Enlightenment Bible*.

18 Although there are some texts in the collection of which there are indeed similar earlier versions, such as the "Pied Piper of Hamelin," their number is relatively small and few of them have played a large part in the success of the fairy tales. When we compare them with other earlier versions, such as those by Charles Perrault or Giambattista Basile, the differences quickly stand out.

19 Basedow, *Methodischer Unterricht der Jugend in der Religion und Sittenlehre der Vernunft nach dem in der Philalethie angegebenen Plane*.

20 On the discovery of childhood in the eighteenth century, see the fundamental study by Ariès, *Centuries of Childhood*; see also Krupp, *Reason's Children*, and Steedman, *Strange Dislocations*.

21 Campe, *Über die früheste Bildung junger Kinderseelen*, 81–82.

22 These specifications have far-reaching consequences for the modern child-centered pedagogy of Campe, Basedow, and Pestalozzi. One consequence is that it is best to educate children largely without guidance, for any attempt at guidance will impress itself on the child primarily as force or violence from without. Another consequence of this wax pedagogy is that any force that acts on the child will also have unintended consequences. Figuratively speaking, every bump on one side of the wax pile results in bumps on other sides; see Pestalozzi, *Tagebuch Pestalozzis über die Erziehung seines Sohnes*, 7–18.

23 On this section, see Breithaupt, "The invention of trauma in German Romanticism." On the unfortunately often neglected Beneke, see Breithaupt, "Homo economicus." On trauma narrative, see Sütterlin, *Poetik der Wunde*.

24 Perhaps the Grimms excluded it from later editions because it was seen as non-German in its origin or because of its violence.

25 Perrault, "Bluebeard."
26 Perrault, "Bluebeard."
27 Bruno Bettelheim provides a psychoanalytic interpretation of this separation and abandonment, as well as of the figure of the evil stepmother. Painting the mother as an evil stepmother makes the separation that is needed for the child's development easier; see Bettelheim, *The Uses of Enchantment*.
28 Thus Grätz, *Das Märchen in der deutschen Aufklärung*, 80. "Jorinde and Joringel" first appeared in 1777 in the autobiography of Jung-Stilling edited by Goethe; see Jung-Stilling, *Henrich Stillings Jugend, Jünglingsjahre, Wanderschaft und häusliches Leben*.
29 Vivasvan Soni has identified the structure of the test as one of the two basic structures of narration. According to Soni, the central feature of test narration is that the question of the protagonist's happiness is suspended until the test is decided. This can occur in narratives of temptation (Jesus in the desert), enduring misfortune (Job), resolving difficult situations, or moral challenges; see Soni, "Trials and tragedies."
30 On Gottsched, see Kiss, "Reinventing the plot." For more on the notion of traumatic injury, see Sütterlin, *Poetik der Wunde*; and Breithaupt, "The invention of trauma in German Romanticism."
31 On tragedies of the seventeenth and eighteenth centuries, see Szondi, *An Essay on the Tragic*.
32 See Pistrol, "Vulnerability."
33 Frevert, *The Politics of Humiliation*.
34 Exceptions include Eriksson and Coultas, "Corpses, maggots, poodles and rats"; as well as the studies cited above, see Heath et al., "Emotional selection in memes," and Stubbersfield et al., "Chicken tumors and a fishy revenge."
35 Here and below, I summarize the findings of Breithaupt et al., "Serial reproduction of narratives preserves emotional appraisals." We also analyzed the resulting data with even more stories in He et al., "Quantifying the retention of emotions across story retellings."
36 On emotions in narratives and their analysis see Breger, "Affects in configuration."
37 For all methods, see Breithaupt et al., "Serial reproduction of narratives preserves emotional appraisals"; see also Mesoudi and Whiten, "The multiple roles of cultural transmission experiments in understanding human cultural evolution"; and Norenzayan et al., "Memory and mystery."
38 For an overview of methods and a helpful discussion, see Elkins, *The Shapes of Stories*.

39 On narratives motivating transmission, see Rimé, "Emotion elicits the social sharing of emotion." For examples of abridgments, see the sometimes crude abridgments of familiar tales that Ray Cashman has collected in Ireland; Cashman, *Packy Jim*. Gordon Allport and Leo Postman called the effect of foreshortening "leveling" in their study of war anecdotes; see Allport and Postman, *The Psychology of Rumor*.

40 For this story and its retellings below, see Breithaupt et al., "Serial reproduction of narratives preserves emotional appraisals."

41 For this story and its retellings, see Breithaupt et al., "Serial reproduction of narratives preserves emotional appraisals."

42 Other studies on disgust conducted with different methods did record an emphasis on disgust. For example, when participants were asked which stories they would retell, they often selected stories of disgust. However, the degree to which disgust was retained was not examined; see Eriksson and Coultas, "Corpses, maggots, poodles and rats"; see also Stubbersfield et al., "Chicken tumors and a fishy revenge."

43 For a study of warnings about drug side effects, see Moussaïd et al., "The amplification of risk in experimental diffusion chains"; see also Jagiello and Hills, "Bad news has wings."

44 Breithaupt et al., "Fact vs. affect in the telephone game."

45 Tobin, *Elements of Surprise*.

46 Yang et al., "Exploring the limits of ChatGPT for query or aspect-based text summarization"; Garrido-Merchán et al., "Simulating H. P. Lovecraft horror literature with the ChatGPT large language model."

47 Kaland et al., "The strange stories test," 80.

48 Brunet-Gouet et al., "Do conversational agents have a theory of mind?"; Kosinski, "Theory of mind may have spontaneously emerged in large language models."

49 Doherty and Perner, "Mental files."

50 For this story and its retellings below, see Breithaupt et al., "Humans create more novelty than ChatGPT when asked to retell a story."

51 For full results, see Breithaupt et al., "Humans create more novelty than ChatGPT when asked to retell a story."

52 Kaup et al., "Processing negated sentences with contradictory predicates"; Hosseini et al., "Understanding by understanding not."

53 Hofstadter, "Gödel, Escher, Bach, and AI."

54 Nabi, "The case for emphasizing discrete emotions in communication research."

55 Antonio Damasio, for example, has suggested that intuitive somatic *markers* may guide our decisions—that is, vague feelings that are based on past

experiences and that can be recalled more quickly than rational calculations; see Damasio, *Descartes' Error.* Similarly, Gerd Gigerenzer speaks of gut *feelings* that we use as basic heuristics to make decisions; see Gigerenzer, *Gut Feelings.*

56 Ranganath, *Why We Remember.* For recall of emotional words, see Bower, "Mood and memory"; and Kensinger and Schacter, "Memory and emotion." For false memory syndrome, see Gleaves et al., "False and recovered memories in the laboratory and clinic"; and Winograd and Neisser, *Affect and Accuracy in Recall.* On schemas and stereotypes, see Wegner and Gray, *The Mind Club;* and Kashima, "Maintaining cultural stereotypes in the serial reproduction of narratives."

57 At least the kind of surprise we tested. Of course, there is also the type of surprise that follows an uncertain outcome when no one knows what will happen next, such as the nail-biter or cliffhanger.

Chapter 3. Emotions as Reward for Narrative Thinking

1 During a field trip to the prison, my students and I were told that McVeigh was its "most prominent guest." For interviews with survivors and relatives, see Madeira, *Killing McVeigh.*

2 Ayano, "Dopamine: Receptors, functions, synthesis, pathways, locations and mental disorders"; Serafini et al., "The mesolimbic dopamine system in chronic pain and associated affective comorbidities"; Leite and Izquierdo, "Generating reward structures on a parameterized distribution of dynamics tasks."

3 Anne Bartsch distinguishes between different ways in which emotional experiences in media and film can be perceived as rewarding. First, according to Bartsch, the experience of emotions in itself is already perceived as positive. In addition, emotions are associated with numerous social and cognitive factors that can be perceived as rewarding, such as parasocial relationships with characters, contemplative experience, and empathetic co-experience; see Bartsch, "Emotional gratification in entertainment experience."

4 I borrow this phrase from Vermeule, *Why Do We Care About Literary Characters?*

5 On blocking empathy, see Breithaupt, *The Dark Sides of Empathy,* chapter 2.

6 For the threat example and elaborations, see Frijda et al., "The duration of affective phenomena or emotions, sentiments and passions"; Frijda, "The laws of emotion," 349. For the triggers that activate empathetic co-

7 Bartsch, "Emotional gratification in entertainment experience."

8 On this argument, see Starr, "Theorizing imagery, aesthetics, and doubly directed states"; see also Scarry, *Dreaming by the Book*.

9 This effect of liking a second listening is well known in music; see Margulis, *On Repeat*.

10 Voss, *Narrative Emotionen*, 38–40; see also Schindler et al., "Measuring aesthetic emotions."

11 For estimates of daydreaming time, see McMillan et al., "Ode to positive constructive daydreaming." See also Mooneyham and Schooler, "The costs and benefits of mind-wandering"; Vago and Zeidan, "The brain on silent"; and Smallwood and Andrews-Hanna, "Not all minds that wander are lost."

12 Dorsch, "Focused daydreaming and mind-wandering," 791.

13 Daydreaming can be associated with many negative emotions for some, such as depressed people, but also with positive and identity-affirming emotions; see Smallwood and Andrews-Hanna, "Not all minds that wander are lost."

14 Kucyi et al., "Individual variability in neural representations of mind-wandering."

15 Hwang et al., "Self-reported expression and experience of triumph across four countries." I thank Ivette Dreyer for the reference.

16 See Soni, "Trials and tragedies."

17 In contrast to the modern miracle stands the medieval miracle, which did not require plausibility, since divine intervention was commonly regarded as real. Rather, the medieval miracle reveals, in exemplary fashion, the power of Christ and thus supports the institution of the church. Accordingly, few people "wonder" about the miracles in medieval texts; see Rüth, "Representing wonder in medieval miracle narratives."

18 See Breithaupt, "Staunen als Belohnung der Neugier."

19 For the "aha" moment, see Topolinski and Reber, "Gaining insight into the 'Aha' experience"; on the rewarding function of curiosity, see Litman, "Curiosity and metacognition"; see also Friston et al., "Active inference and epistemic value."

20 On wonder as a drive toward knowledge, see Matuschek, *Über das Staunen*; Daston, "Curiosity in early modern science"; on the "absolute limit" of the near-impossible experience, see Gess and Schnyder, "Staunen als Grenzphänomen," 8.

21 See Scherer et al., *Appraisal Processes in Emotion*.

22 On the role of motivations in emotions, see Roseman et al., "Appraisals of emotion-eliciting events."

23 See, for example, Reinhard M. Möller, who argues that astonishment as a goal of aesthetic experience promises both an experience of the self and of the other; Möller, "Ästhetiken gegenstandsbezogenen Staunens im 18. Jahrhundert."

24 Miall, "Anticipation and feeling in literary response," 276.

25 On the inner life of characters, see Zunshine, *Why We Read Fiction;* see also Jannidis, *Figur und Person,* 185–195. For suspense, see Carroll, *The Philosophy of Horror;* see also Ortony et al., *The Cognitive Structure of Emotions.*

26 Flesch, *Comeuppance.*

27 Haidt, *The Righteous Mind.* Peter DeScioli and Robert Kurzban argue that morality allows people to come to the same conclusion about conflicts ("bystander-coordination"), thereby avoiding splits within larger groups; see DeScioli and Kurzban, "A solution to the mysteries of morality."

28 Jolles, *Simple Forms.*

29 Flesch, *Comeuppance.*

30 See Gill and Getty, "On shifting the blame to humanity."

31 Sen, *The Idea of Justice,* introduction.

32 See Hanich et al., "Why we like to watch sad films."

33 See Menninghaus et al., "Towards a psychological construct of being moved"; see also Cova and Deonna, "Being moved."

34 Kelley, "A modern Cinderella."

35 See Breithaupt, *The Dark Sides of Empathy.*

36 See Boitani, *Anagnorisis,* 12.

37 See Cave, *Recognitions.*

38 Homer, *The Odyssey,* 22, 39/40.

39 Albrecht Koschorke has traced the career of the vocabulary of fluid circulations and movements of the body based on the discussions of the eighteenth century, the century in which modern aesthetics was founded; see Koschorke, *Körperströme und Schriftverkehr.*

40 Geulen, "Anagnorisis statt Identifikation," 225, 226; Moretti, *Signs Taken for Wonders.*

41 I thank Christoph Irmscher for this idea.

42 See Cova and Deonna, "Being moved."

43 It is quite possible that scenes of recognition and scenes of return play a greater role in my sensibilities than they do for other people. I don't want to put myself on the Freudian couch now, but there is a parallel here for me in that I too experienced the shock of my father's absence when I was ten years old. At that time, my father, a diplomat for the West German

Department of Defense, did not return from an excursion and was missing for a few days in a wilderness, where military helicopters searched for him. Unlike my mother's father, however, he did not return but was found dead under mysterious circumstances.

44 Schmid, "Eventfulness and repetitiveness."

45 Žižek, *Event*, 12. Furthermore, Žižek argues that an event is an "effect that seems to exceed its causes" (6).

46 Goethe, *Wilhelm Meister's Apprenticeship and Travel*, I, 3.

47 Breithaupt et al., "Fact vs. affect in the telephone game."

48 Sometimes surprise is classified as negative (Topolinski and Strack, "Corrugator activity confirms immediate negative affect in surprise"); at times as neutral (Reisenzein, "Exploring the strength of association between the components of emotion syndromes"); and at times as positive (Fontaine et al., "The world of emotions is not two-dimensional"). See also Noordewier et al., "The temporal dynamics of surprise."

49 Knutson and Cooper, "The lure of the unknown"; Kidd and Hayden, "The psychology and neuroscience of curiosity."

50 Dubourg and Baumard, "Why imaginary worlds?"

51 Goffman, *Presentations of Self in Everyday Life*, and Goffman, *Stigma*.

52 Lewis, "Self-conscious emotions."

53 See "The rhetoric of temporality" by Paul de Man, originally published in 1969 and reprinted in the collection of his essays *Blindness and Insight* (1983).

54 Academic research's neglect of love as an emotion is documented, for example, in the transition from the third to the fourth edition of the *Handbook of Emotions* between 2008 and 2016. In the earlier edition, there is no entry with the keyword *love*, but in 2016 there is an article on the topic, to which we will refer shortly. On the notion of love as a narrative rather than an emotion, see Ekman, "Basic emotions": Ekman justifies this exclusion with the temporal duration of love, and, of particular interest to us, with the fact that love represents an "emotional plot" (55). For Ekman, then, love is not a *basic emotion* precisely because it is narrative. I should indicate here that I consider this definitional demarcation absurd and a misapplication and outgrowth of the appraisal theory of emotions. (Perhaps there is also an only half-reflected male prejudice behind Ekman's theory about what is to be considered an emotion: short and quick, then it's over. All this is not to deny Ekman's considerable achievement.)

55 Fredrickson, "Love: Positivity resonance as a fresh, evidence-based perspective on an age-old topic," 848 and 852.

56 Eva Illouz also emphasizes this temporal element of love; see Illouz, *The End of Love*. For Illouz, love and sex are an expression of freedom expressed

by means of participation, choice, and consumption. Illouz suggests that
we experience love and sex as freedom insofar as they contain the choice
to participate or not. This choice also allows us to grasp the freedom of
love as narrative. Indeed, many of the case histories Illouz presents are
attempts by the narrators to explain—and narrate—what actually happened after the fact. This means, from my perspective, that this form of
love as freedom of consumption exists only as narrative, for it is only in
narrative that the freedom of choice can be conveyed and perceived. The
narrator's reward, then, is to experience the previously uncertain, fragile,
and momentary act of love as a freedom.

57 See also Gottschall, *The Story Paradox.*

58 Freud readers will certainly think of the constellation of the Oedipus
complex in which identification with the father makes him a rival.

59 For all their differences, this assumption is shared by a wide range of
thinkers from Georg Simmel and Niklas Luhmann to Michel Foucault.

60 Foucault, *Confessions of the Flesh.*

61 Schulz and Hübner, "Mittelalter."

62 Campbell, *The Hero with a Thousand Faces.* Emmanuel Levinas opposed
this schema of the self-enriching returnee with the figure of Moses. Moses
departs without experiencing return and arrival; see Lévinas, "The trace
of the other"; see also Kolmar, *Grenzbeschreibungen.*

63 On Casanova's apparent selflessness as a trap, see Masters, *Casanova*, 61.
Casanova's assertions of willingness are gross beautifications. While he
repeatedly holds to the trope of shared pleasure, he also regularly uses
it only as a retrospective justification of prior violence. Casanova tends
to liken "any woman ravished to a woman willing," according to Brin,
"'Triompher par la force,'" 23.

64 On the often neglected sense of smell, see Barwich, *Smellosophy.*

65 Decety, "Human empathy."

66 Damasio, *Descartes' Error*; Gigerenzer and Todd, *Simple Heuristics That
Make Us Smart.*

67 On overcoming negative experiences through *redemptive narratives*, see
the work of Dan P. McAdams, such as McAdams and McLean, "Narrative identity"; Rogers et al., "Seeing your life story as a Hero's Journey
increases meaning in life." On the potential selective benefits of narrative
thinking, see Bietti et al., "Storytelling as adaptive collective sensemaking"; Boyd, *On the Origin of Stories.*

68 Voss, *Narrative Emotions.*

69 Ngai, *Ugly Feelings*, 12, 18.

70 Ngai, *Ugly Feelings*, 209–211, 21.

71 This expression is to be understood differently here than the critical feelings that Rolf Reber describes. Reber emphasizes the possibility for everyone to learn from his or her feelings; see Reber, *Critical Feeling*. Ngai, on the other hand, sees the term *critical* in the tradition of Adorno as a counterpoint to the world as is.

72 Nussbaum, *Political Emotions*, 15.

Chapter 4. Narratives as Response to a Crisis

1 This idea also stands at the foundation of medical humanities; see Charon, *Narrative Medicine*.

2 Koschorke, *Fact and Fiction*; see Flesch, *Comeuppance*, for the reward and punishment structure; on the collective creation of meaning, see Ächtler, "Was ist ein Narrativ?" In *Sapiens*, Yuval Harari combines the collective narrative's adaptability with Robin Dunbar's argument that gossip creates cohesion in groups; see Dunbar, *Grooming, Gossip, and the Evolution of Language*.

3 Schulze, *Germany*.

4 For more detail on these and the following considerations, see Breithaupt, "Rituals of trauma."

5 Chomsky, *9-11*.

6 See Bruner, *Making Stories*.

7 McAdams and McLean, "Narrative identity."

8 See Schmid, "Eventfulness and repetitiveness."

9 Here we should again think of Sianne Ngai's suggestion that ugly emotions arise when the classical or cathartic flow of a narrative is interrupted; see Ngai, *Ugly Feelings*.

10 Rogers et al., "Seeing your life story as a Hero's Journey increases meaning in life." For the hero's journey, see Campbell, *The Hero with a Thousand Faces*.

11 On the connections between narration and institutions, see the study by Koschorke, *Fact and Fiction*.

12 We may consider René Girard's analysis here. In *Violence and the Sacred*, Girard argues that the voice of authority, such as the voice of law, overrides ongoing blood feuds between clans.

13 Genette, *Narrative Discourse*.

14 The idea of this hope for immunity by means of repetition is also explored by Türk, *Die Immunität der Literatur*.

15 See Babington, *Shell-Shock*.

16 This insight is initially found primarily in works of German Romanticism.

On the distinctly interesting literary discovery of trauma, see Sütterlin, *Poetik der Wunde*, and Breithaupt, "The invention of trauma in German Romanticism."

17 On Balzac, see Felman, *Writing and Madness*.
18 Goethe, *Wilhelm Meister's Apprenticeship*.
19 Türk, *Die Immunität der Literatur*.
20 Lauer, "Das Erdbeben von Lissabon."

Chapter 5. Narrative Characters

1 Martínez, "Figuration," 145. See also Jannidis, *Figur und Person*. The difference between real person and figure here does not correspond to the difference between "actor" and "character" used by Mieke Bal, insofar as "actor" is already a constellation in narratives, even if only vaguely recognized; scc Bal, *Narratology*, 114–132.

2 On tracking, see Flesch, *Comeuppance*, 17–21.

3 The nobility, for example, were granted immunity from tax payments and thus were elevated above others in perpetuity; see Türk, *Die Immunität der Literatur*.

4 Breithaupt, *Der Ich-Effekt des Geldes*, introduction.

5 Perner et al., "Mental files and belief."

6 Doherty and Perner, "Mental files."

7 An interesting question is whether this form of interactive fiction generates more or less empathy for the characters in readers/players. In particular, it has not been examined how empathy is "divided" among multiple characters. We explored this question by comparing texts in which readers either had no choice (traditional fiction) or had to make choices for two characters, each of which also affected the other character. The result was that interactive fiction moved many readers to a clear preference for one or the other character. They then had more empathy for this character but neglected the other character. In this case, many participants in our experiment made choices that could even harm the acting character but helped their preferred character. In traditional fiction, by contrast, empathy was less polarized; see Lagrange et al., "Pawn-playing and biased empathy."

8 On *Detroit: Become Human*, see Lagrange, "Individualized communal experience."

9 For an overview, see Gao et al., "Examining whether adults with autism spectrum disorder encounter multiple problems in theory of mind."

10 One should keep in mind that the notion of play and playfulness is clearly context-dependent; see Allen and Bekoff, *Species of Mind*.

11 Gardner and Daw, "Tulpamancy." Questions arise here that echo Elizabeth Schechter's doubts about the unity of consciousness; see Schechter, "The unity of consciousness." Schechter is working on a manuscript on tulpamancy, and I was fortunate to hear her early reflections as a lecture.

12 See Luhrmann et al., "Learning to discern the voices of gods, spirits, tulpas, and the dead."

13 Princefirestorm, "Switching and co-fronting."

14 For an examination of the positive effects of inner speech, see Fernyhough and Borghi, "Inner speech as language process and cognitive tool."

15 Cook, *Building Character.*

16 Breithaupt, *Der Ich-Effekt des Geldes;* Simmel, *Philosophie des Geldes;* Butler, *Giving an Account of Oneself;* see also Breithaupt, *Kultur der Ausrede.*

17 See Bruner, *Making Stories.*

18 Bruner, *Making Stories.*

19 McAdams and McLean. "Narrative identity"; McAdams, "'First we invented stories, then they changed us.'"

20 Strawson, "Against narrativity."

21 Gigerenzer and Todd, "Précis of simple heuristics that make us smart."

22 Lagrange et al., "Pawn-playing and biased empathy"; Mounk, *The Identity Trap.*

23 In *The Mind Club,* Wegner and Gray examine the dyadic patterns by which we sort the relationships between characters and persons in order to orient ourselves in the social world with beings with their own "mind" or spirit. On sensitivity to victims, see Hunt, *Inventing Human Rights.* Steven Pinker has collected impressive data to support his thesis that our world is moving in positive prosocial directions where violence is increasingly becoming the exception; see Pinker, *The Better Angels of Our Nature,* and Pinker, *Enlightenment Now.*

24 Breithaupt, *The Dark Sides of Empathy,* chapter 3.

25 Beck, "Thinking and depression."

26 See Bollen et al., "Historical language records reveal a surge of cognitive distortions in recent decades." This trend, it should be noted, feeds from uses in works of fiction as well as nonfiction found in the vast Google corpus.

Chapter 6. Multiversional Reality, Multiversional Narratives

1 This tension is an aspect of what Marco Caracciolo describes as the experientiality of narrative; see Caracciolo, *The Experientiality of Narrative.*

2 Friston et al., "A free-energy principle for the brain"; Friston, "The free-energy principle: A unified brain theory."

3 Clark, *The Experience Machine.*

4 Hohwy, *The Predictive Mind*; Clark, *Surfing Uncertainty*; Knill and Pouget, "The Bayesian brain."

5 Hutchinson and Feldman Barrett, "The power of predictions," 281.

6 Friston, "The free-energy principle."

7 Hutchinson and Feldman Barrett, "The power of predictions."

8 An exception is Yon et al., "Beliefs and desires in the predictive brain."

9 See Breithaupt, *Kultur der Ausrede*, chapter 1; on the following remarks, see also Hiskes et al., "Multiversionality."

10 See also Fletcher, *Storythinking.*

11 Ortony et al., *The Cognitive Structure of Emotions*; Carroll, "The paradox of suspense."

12 On suspense and positivity, see Bálint et al., "The effect of suspense structure on felt suspense and narrative absorption in literature and film."

13 On the reinforcements of the feeling of presence in general, see Gumbrecht, *Production of Presence.*

14 For an introduction and overview, see Smuts, "The paradox of suspense"; see also Daniel and Katz, "Spoilers affect the enjoyment of television episodes but not short stories."

15 See Margulis, *On Repeat.*

16 Hohwy, *The Predictive Mind*; on the therapeutic effect of narrative, see, for example, Brockington et al., "Storytelling increases oxytocin and positive emotions and decreases cortisol and pain in hospitalized children"; Bezdek et al., "'Run for it!': Viewers' participatory responses to film narratives."

17 See Kukkonen, *Probability Designs.*

18 See, for example, Gerrig and Wenzel, "The role of inferences in narrative experiences."

19 Brooks, *Reading for the Plot.*

20 See Kukkonen, *Probability Designs.*

21 For the situation models, see Zwaan et al., "The construction of situation models in narrative comprehension." New variants of the situation model emphasize the limits of situation spaces; it is usually assumed that one is in only one event space at a time; see Radvansky and Zacks, *Event Cognition.*

22 A horrific case of redirecting empathy has been discussed by Hannah Arendt, who refers to the advice Himmler gave to Nazi executioners. Instead of thinking of the victims of their atrocities, the executioners were supposed to feel sorry for themselves, to have self-empathy, because they

were doing such hard duty; see Breithaupt, *The Dark Sides of Empathy*, chapter 2.

23 Lagrange, "Individualized communal experience."

Chapter 7. Evolution of the Narrative Brain

1 This is also true for emotional empathy or affect sharing according to the "perception action model," which suggests that we use similar routines when we observe an action and when we ourselves perform the action; see Decety, "Human empathy."

2 Green et al., "Emotion and transportation into fact and fiction"; Martínez, "Storyworld possible selves and the phenomenon of narrative immersion."

3 Here are some of my reference points. David Chalmers has coined the phrase the "hard problem"; see *The Conscious Mind*. Douglas Hofstadter has helped me consider self-references; see *I Am a Strange Loop*. Another starting point comes from Antonio Damasio, who places feelings at the central hinge between inner experience and environment, insofar as feelings provide people with inner feedback about the external situation; see *Feeling and Knowing*.

4 On the complex of consciousness as sensory perception, see Armstrong, *The Nature of Mind*, 55–67.

5 Gumbrecht, *Productions of Presence*, and Christner, *How Literature Produces Presence*.

6 On the distinction between different forms of empathy, see Batson, "These things called empathy."

7 I refer to archaeological findings of musical instruments and assumptions about the origin of language. Neanderthals may have practiced cave painting more than 64,000 years ago; see Hoffmann et al., "U-Th dating of carbonate crusts reveals Neandertal origin of Iberian cave art." Ochre was apparently mined in Kenya as early as 285,000 years ago; see Ackerman, "Prehistoric arts and crafts."

8 In the context of collective experiences with music concerts, see Merrill et al., "The aesthetic experience of live concerts."

9 See Dunbar, *Grooming, Gossip, and the Evolution of Language*; Bietti et al., "Storytelling as adaptive collective sensemaking."

10 The classic model goes back to Atkinson and Shiffrin, "Human memory."

11 See the work of Husserl's student Schapp, *In Geschichten verstrickt*.

12 This was already suggested by Dennett in *Consciousness Explained*.

13 On the conception of transportation, see Green, "Transportation into narrative worlds."

14 Nagel, "What is it like to be a bat?"

15 In this respect, it is no wonder that we seem to have developed not just one but several empathy-related mechanisms. For example, it is now considered well established among neuroscientists that we have at least two different brain routines for understanding the internal states of others (theory of mind) and for copying their emotional states into our routines (affect sharing); see Kanske, "The social mind."

16 See Elam, *The Semiotics of Theatre and Drama.*

17 Call and Tomasello, "Does the chimpanzee have a theory of mind?"; Tomasello, *The Origins of Human Communication.*

18 Kobayashi and Kohshima, "Unique morphology of the human eye and its adaptive meaning."

19 Tomasello, *The Origins of Human Communication.*

20 In regard to misdirection, see Rothstein, *The Shape of Difficulty.* On manipulation in storytelling, see Breithaupt, *Kultur der Ausrede.*

21 The Swiss zoologist Hans Kummer was the first to document this behavior; see Kummer, *Social Organization of Hamadryas Baboons.*

22 Merrill et al., "The aesthetic experience of live concerts."

23 Schechner, *Between Theater and Anthropology.* For more on the mental event of the audience, see Schmid, *Mentale Ereignisse.*

24 On side-taking, see Woodward et al., "Spontaneous side-taking."

25 A first mimetic and pre-linguistic phase of narration is assumed by Porter Abbott in "The evolutionary origins of the storied mind."

26 Slocombe and Zuberbühler, "Functionally referential communication in a chimpanzee."

27 Fitch, *Origins of Language.*

28 It is possible that our episodic memory capacity has distinct communicative functions. Most researchers emphasize the benefits of planning (good memory allows planning of future actions). However, episodic memory and communication might also serve to persuade others to believe the speaker and agree with his or her conclusions. According to Johannes Mahr and Gergely Csibra, this includes, for example, argumentative forms of generalization (observing someone committing a moral transgression and then communicating this action can persuade others to collectively track the other). If the communicated episode is successful, this leads, for instance, to commitment and cooperation; see Mahr and Csibra, "Why do we remember?"

29 On casting as a cognitive performance, see Cook, *Building Character.* In

regard to "pretend play," see Tooby and Cosmides, "Does beauty build adapted minds?"

30 Mellmann, "Evolutionary proto-forms of literary behavior"; Warburton, "Of meanings and movements."

31 Heidegger, *Plato*.

32 Decoupling thus becomes the cognitive hinge that distinguishes different forms of representation; see Cosmides and Tooby, "Adaptations for decoupling and metarepresentation."

33 See Cosmides and Tooby, "Adaptations for decoupling and metarepresentation."

Epilogue

1 Among prominent theorists of narratology whose works operate in the realm of this model, I include Seymor Chatman, Wolfgang Iser, Jurij Lotman, Tzvetan Todorov, and Wolf Schmid, as well as some works by Monika Fludernik and James Phelan. It is also worth recalling, for inspiration, a work by Walter Benjamin, "The storyteller," in which Benjamin shows how even the simplest events in narratives always have more than one interpretation and are thus at the same time simple and highly complex.

2 See Žižek, *Event*.

3 See Hamilton and Breithaupt, "These things called event."

4 Benjamin, "The storyteller."

5 See also Iser, *The Act of Reading*.

6 Genette, *Narrative Discourse*.

7 Angus Fletcher suggests that story thinking trains perspective switching. I agree but would also caution that stories also provide resistance against such switching, as we grow affectively comfortable in each perspective; see Fletcher, *Storythinking*.

8 Breithaupt, *The Dark Sides of Empathy*, chapter 3.

9 In this sense, Lisa Zunshine shows how reading narratives involves theory of mind; Zunshine, *Why We Read Fiction*.

10 Gottschall, *The Story Paradox*, 8–11, 12; see also Lazer et al., "The science of fake news."

11 Gottschall, for example, calculates how many hours a day we are enmeshed in stories, even if not every video game has to be called a story; see Gottschall, *The Story Paradox*, 19–21.

12 I am not as positive as Chalmers about these technologies; see Chalmers, *Reality+*.

13 "For sale: Baby shoes, never worn." For this anecdote, see Gottschall, *The Story Paradox*, 53–54.
14 Rosa, *Social Acceleration.*
15 Dunbar, *Grooming, Gossip, and the Evolution of Language.*

BIBLIOGRAPHY

Abbott, H. Porter. "The evolutionary origins of the storied mind: Modeling the prehistory of narrative consciousness and its discontents." *Narrative* 8, no. 3 (2000): 247–256.

Acerbi, Alberto, and Joseph M. Stubbersfield. "Large language models show human-like content biases in transmission chain experiments." *Proceedings of the National Academy of Sciences* 120, no. 44 (2023): 2313790120.

Ächtler, Norman. "Was ist ein Narrativ? Begriffsgeschichtliche Überlegungen anlässlich der aktuellen Europa-Debatte." *KulturPoetik* 14, no. 2 (2014): 244–268.

Ackerman, Sandra J. "Prehistoric arts and crafts." *American Scientist* 106, no. 1 (2018): 8–11.

Allen, Colin, and Marc Bekoff. *Species of Mind: The Philosophy and Biology of Cognitive Ethology.* Cambridge: MIT Press, 1999.

Allport, Gordon, and Leo Postman. *The Psychology of Rumor.* Oxford: Henry Holt, 1947.

Ariès, Philippe. *Centuries of Childhood: A Social History of Family Life.* New York: Vintage, 1962.

Armstrong, David M. *The Nature of Mind.* Ithaca: Cornell University Press, 1981.

Armstrong, Paul. *Stories and the Brain: The Neuroscience of Narrative.* Baltimore: Johns Hopkins University Press, 2020.

Atkinson, Richard C., and Richard M. Shiffrin. "Human memory: Proposed system and its control processes." *Psychology of Learning and Motivation* (1968): 89–195.

Ayano, Getinet. "Dopamine: Receptors, functions, synthesis, pathways, locations and mental disorders: Review of literatures." *Journal of Mental Disorders and Treatment* 2, no. 120 (2016): 1–4.

Babington, Anthony. *Shell-Shock: A History of the Changing Attitudes to War Neurosis.* London: Leo Cooper, 1997.

Bal, Mieke. *Narratology: Introduction to the Theory of Narrative.* Toronto: University of Toronto Press, 2002.

Bálint, Katalin, Moniek M. Kuijpers, Miruna M. Doicaru, Frank Hakemulder, and Ed S. Tan. "The effect of suspense structure on felt suspense and narrative absorption in literature and film." *Narrative Absorption* 27, no. 77 (2017).

Baron-Cohen, Simon, Sally Wheelwright, Jacqueline Hill, Yogini Raste, and Ian Plumb. "The 'Reading the Mind in the Eyes' Test revised version: A study with normal adults, and adults with Asperger syndrome or high-functioning autism." *Journal of Child Psychology and Psychiatry and Allied Disciplines* 42, no. 2 (2001): 241–251.

Bartlett, Frederic. *Remembering: A Study in Experimental and Social Psychology.* Cambridge: Cambridge University Press, 1932.

Bartsch, Anne. "Emotional gratification in entertainment experience: Why viewers of movies and television series find it rewarding to experience emotions." *Media Psychology* 15, no. 3 (2012): 267–302.

Barwich, Ann-Sophie. *Smellosophy.* Cambridge: Harvard University Press, 2020.

Basedow, Johann Bernhard. *Methodischer Unterricht der Jugend in der Religion und Sittenlehre der Vernunft nach dem in der Philalethie angegebenen Plane.* Altona, 1764.

Batson, Daniel. "These things called empathy: Eight related but distinct phenomena." In *The Social Neuroscience of Empathy,* edited by Jean Decety, 3–15. Cambridge: MIT Press, 2009.

Beck, Aaron T. "Thinking and depression: I. Idiosyncratic content and cognitive distortions." *Archives of General Psychiatry* 9, no. 4 (1963): 324–333.

Benjamin, Walter. "The storyteller: Reflections on the works of Nikolai Leskov." In *Illuminations,* translated by Harry Zorn, 83–107. New York: Harvard University and Harcourt, 1968.

Bergman, Erik T., and Henry L. Roediger. "Can Bartlett's repeated reproduction experiments be replicated?" *Memory and Cognition* 27, no. 6 (1999): 937–947.

Bettelheim, Bruno. *The Uses of Enchantment: The Meaning and Importance of Fairy Tales.* New York: Vintage, 1989.

Bezdek, Matthew A., Jeffrey E. Foy, and Richard J. Gerrig. "'Run for it!': Viewers' participatory responses to film narratives." *Psychology of Aesthetics, Creativity, and the Arts* 7, no. 4 (2013): 1–8.

Bietti, Lucas M., Ottilie Tilston, and Adrian Bangerter. "Storytelling as adaptive collective sensemaking." *Topics in Cognitive Science* 11, no. 4 (2019): 710–732.

Bloom, Harold. *Shakespeare: The Invention of the Human.* New York: Riverhead, 1998.

Boddice, Rob. *The History of Emotions.* 2nd ed. Manchester: Manchester University Press, 2024.

Boitani, Piero. *Anagnorisis: Scenes and Themes of Recognition and Revelation in Western Literature*. Leiden: Brill, 2021.

Bollen, Johan, Marijn Ten Thij, Fritz Breithaupt, Alexander Barron, Lauren A. Rutter, Lorenzo Lorenzo-Luaces, and Marten Scheffer. "Historical language records reveal a surge of cognitive distortions in recent decades." *Proceedings of the National Academy of Sciences* 118, no. 30 (2021): e2102061118.

Bornstein, Marc H., Kay Ferdinandsen, and Charles G. Gross. "Perception of symmetry in infancy." *Developmental Psychology* 17, no. 1 (1981): 82.

Bower, Gordon. "Mood and memory." *American Psychologist* 36, no. 2 (1981): 129.

Boyd, Brian. *On the Origins of Stories: Evolution, Cognition, and Fiction*. Cambridge: Harvard University Press, 2009.

Breger, Claudia. "Affects in configuration: A new approach to narrative worldmaking." *Narrative* 25, no. 2 (2017): 227–251.

Breithaupt, Fritz. *The Dark Sides of Empathy*. Ithaca: Cornell University Press, 2019.

———. "Staunen als Belohnung der Neugier: Wunder, Überraschung und Frage in narrativer Hinsicht." In *Poetik des Wunders*, edited by Myreille Schneider, Nicola Gess, and Hugues Marchal, 37–49. Munich: Wilhelm Fink Verlag, 2019.

———. *Kultur der Ausrede*. Berlin: Suhrkamp, 2012.

———. *Kultur der Empathie*. Frankfurt: Suhrkamp, 2009.

———. *Der Ich-Effekt des Geldes: Zur Geschichte einer Legitimationsfigur*. Frankfurt: Fischer, 2008.

———. "Homo economicus: The rhetoric of currency and a case of nineteenth century psychology." *Nineteenth-Century Prose* 32, 1 (2005): 6–26.

———. "The invention of trauma in German Romanticism." *Critical Inquiry* 32, no. 1 (2005): 77–101.

———. "Rituals of trauma: How the media fabricated 9/11." In *Media Representations of September 11*, edited by Steven Chermak, Frankie Y. Bailey, and Michelle Brown, 67–81. Westport, Conn.: Praeger, 2003.

Breithaupt, Fritz, Binyan Li, and John K. Kruschke. "Serial reproduction of narratives preserves emotional appraisals." *Cognition and Emotion* 36, no. 4 (2022): 581–601.

Breithaupt, Fritz, Binyan Li, Torrin M. Liddell, Eleanor B. Schille-Hudson, and Sarah Whaley. "Fact vs. affect in the telephone game: All levels of surprise are retold with high accuracy, even independently of facts." *Frontiers in Psychology* 9 (2018): 2210.

Breithaupt, Fritz, Ege Otonen, Devin Wright, John Kruschke, Ying Li, and

Yiyan Tan. "Humans create more novelty than ChatGPT when asked to retell a story." *Scientific Reports* 14 (2024): 875.

Brigard, Felipe de. "Is memory for remembering? Recollection as a form of episodic hypothetical thinking." *Synthese* 191, no. 2 (2014): 155–185.

Brin, Raphaëlle. "'Triompher par la force': Sexual violence and its representation in Casanova's *History of My Life*." In *Casanova in the Enlightenment*, edited by Malina Stefanovska, 17–34. Toronto: University of Toronto Press, 2021.

Brockington, Guilherme, Ana Paula Gomes Moreira, Maria Stephani Buso, Sérgio Gomes da Silva, Edgar Altszyler, Ronald Fischer, and Jorge Moll. "Storytelling increases oxytocin and positive emotions and decreases cortisol and pain in hospitalized children." *Proceedings of the National Academy of Sciences* 118, no. 22 (2021): e2018409118.

Brooks, Peter. *Reading for the Plot: Design and Intention in Narrative*. Cambridge: Harvard University Press, 1992.

Bruner, Jerome. *Making Stories: Law, Literature, Life*. Cambridge: Harvard University Press, 2003.

———. *Actual Minds, Possible Words*. Cambridge: Harvard University Press, 1986.

Brunet-Gouet, Eric, Nathan Vidal, and Paul Roux. "Do conversational agents have a theory of mind? A single case study of ChatGPT with the Hinting, False Beliefs and False Photographs, and Strange Stories paradigms." November 15, 2023, https://hal.science/hal-03991530v2.

Butler, Judith. *Giving an Account of Oneself*. New York: Fordham University Press, 2009.

Call, Josep, and Michael Tomasello. "Does the chimpanzee have a theory of mind? 30 years later." *Trends in Cognitive Sciences* 12, no. 5 (2008): 187–192.

Campbell, Joseph. *The Hero with a Thousand Faces* [1949]. Princeton: Princeton University Press, 1968.

Campe, Joachim Heinrich. *Über die früheste Bildung junger Kinderseelen*, edited by Brigitte H. E. Nestroj. Frankfurt: Ullstein, 1985.

Caracciolo, Marco. *The Experientiality of Narrative: An Enactivist Approach*. Berlin: De Gruyter, 2014.

Carroll, Noël. "The paradox of suspense." In *Suspense: Conceptualizations, Theoretical Analyses, and Empirical Explorations*, edited by Peter Vorderer, Hans Jürgen Wulff, and Mike Friedrichsen, 81–102. London: Routledge, 2013.

———. *The Philosophy of Horror: Or, Paradoxes of the Heart*. London: Routledge, 2003.

Cashman, Ray. *Packy Jim: Folklore and Worldview on the Irish Border*. Madison: University of Wisconsin Press, 2016.

Cave, Terence. *Recognitions: A Study in Poetics*. Oxford: Oxford University Press, 1988.

Chalmers, David J. *Reality+: Virtual Worlds and the Problems of Philosophy*. New York: W.W. Norton, 2022.

———. *The Conscious Mind: In Search of a Fundamental Theory*. Oxford: Oxford University Press, 1996.

Chang, Claire H. C., Christina Lazaridi, Yaara Yeshurun, Kenneth A. Norman, and Uri Hasson. "Relating the past with the present: Information integration and segregation during ongoing narrative processing." *Journal of Cognitive Neuroscience* 33, no. 6 (2021): 1106–1128.

Chang, Claire H. C., Samuel A. Nastase, and Uri Hasson. "Information flow across the cortical timescale hierarchy during narrative construction." *Proceedings of the National Academy of Sciences* 119, no. 51 (2022): e2209307119.

Charon, Rita. *Narrative Medicine: Honoring the Stories of Illness*. Oxford: Oxford University Press, 2008.

Chomsky, Noam. *9-11*. New York: Seven Stories, 2001.

Christner, Bettina. *How Literature Produces Presence*. Indiana University ProQuest Dissertations Publishing (2022): 29318654.

Clark, Andy. *The Experience Machine: How Our Minds Predict and Shape Reality*. New York: Pantheon, 2023.

———. *Surfing Uncertainty: Prediction, Action, and the Embodied Mind*. Oxford: Oxford University Press, 2015.

Clark, Andy, and David Chalmers. "The extended mind." *Analysis* 58, no. 1 (1998): 7–19.

Cook, Amy. *Building Character: The Art and Science of Casting*. Ann Arbor: University of Michigan Press, 2018.

Cosmides, Leda, and John Tooby. "Adaptations for decoupling and metarepresentation." In *Metarepresentations: A Multidisciplinary Perspective*, edited by Dan Sperber, 53–115. Oxford: Oxford University Press, 2000.

Cova, Florian, and Julien A. Deonna. "Being moved." *Philosophical Studies* 169, no. 3 (2014): 447–466.

Crystal, Jonathan D. "Elements of episodic-like memory in animal models." *Behavioural Processes* 80, no. 3 (2009): 269–277.

Damasio, Antonio. *Feeling and Knowing: Making Minds Conscious*. New York: Pantheon, 2021.

———. *Descartes' Error*. New York: Random House, 2006.

Daniel, Thomas A., and Jeffrey S. Katz. "Spoilers affect the enjoyment of television episodes but not short stories." *Psychological Reports* 122, no. 5 (2019): 1794–1807.

Daston, Lorraine. "Curiosity in early modern science." *Word and Image* 11, no. 4 (1995): 391–404.

Dawkins, Richard. *The Selfish Gene*. Oxford: Oxford University Press, 2016.

Deacon, Terrence William. *The Symbolic Species: The Co-Evolution of Language and the Brain*. New York, London: Norton, 1998.

Decety, Jean. "Human empathy." *Japanese Journal of Neuropsychology* 22, no. 1 (2006): 11–33.

Dennett, Daniel. *Consciousness Explained* [1991]. New York: Little, Brown, 2017.

DeScioli, Peter, and Robert Kurzban. "A solution to the mysteries of morality." *Psychological Bulletin* 139, no. 2 (2013): 477.

Doherty, Martin J., and Josef Perner. "Mental files: Developmental integration of dual naming and theory of mind." *Developmental Review* 56 (2020): 100909.

Dominey, Peter Ford. "Narrative event segmentation in the cortical reservoir." *PLoS Computational Biology* 17, no. 10 (2021): e1008993.

Dorsch, Fabian. "Focused daydreaming and mind-wandering." *Review of Philosophy and Psychology* 6, no. 4 (2015): 791–813.

Dubourg, Edgar, and Nicolas Baumard. "Why imaginary worlds? The psychological foundations and cultural evolution of fictions with imaginary worlds." *Behavioral and Brain Sciences* 45 (2022): e276.

Dudukovic, Nicole M., Elizabeth J. Marsh, and Barbara Tversky. "Telling a story or telling it straight: The effects of entertaining versus accurate retellings on memory." *Applied Cognitive Psychology* 18, no. 2 (2004): 125–143.

Dunbar, Robin. *Grooming, Gossip, and the Evolution of Language*. Cambridge: Harvard University Press, 1996.

Ekman, Paul. "Basic emotions." *Handbook of Cognition and Emotion* 98 (1999): 45–60.

Elam, Keir. *The Semiotics of Theatre and Drama* [1980]. London: Routledge, 2005.

Elkins, Katherine. *The Shapes of Stories: Sentiment Analysis for Narrative*. Cambridge: Cambridge University Press, 2022.

Eriksson, Kimmo, and Julie C. Coultas. "Corpses, maggots, poodles and rats: Emotional selection operating in three phases of cultural transmission of urban legends." *Journal of Cognition and Culture* 14, no. 1–2 (2014): 1–26.

Felman, Shoshana. *Writing and Madness: Literature/Philosophy/Psychoanalysis*. Ithaca: Cornell University Press, 1985.

Fernyhough, Charles, and Anna M. Borghi. "Inner speech as language process and cognitive tool." *Trends in Cognitive Sciences* 27, no. 12 (2023): 1180–1193.

Fitch, Tecumseh. *Origins of Language*. Cambridge: Cambridge University Press, 2009.

Fleming, Paul. *Exemplarity and Mediocrity: The Art of the Average from Bourgeois Tragedy to Realism*. Palo Alto: Stanford University Press, 2009.

Flesch, William. *Comeuppance: Costly Signaling, Altruistic Punishment, and Other Biological Components of Fiction*. Cambridge: Harvard University Press, 2007.

Fletcher, Angus. *Storythinking: The New Science of Narrative Intelligence*. New York: Columbia University Press, 2023.

Fludernik, Monika. *Towards a 'Natural' Narratology*. London: Routledge, 2002.

Fontaine, Johnny, Klaus R. Scherer, Etienne B. Roesch, and Phoebe C. Ellsworth. "The world of emotions is not two-dimensional." *Psychological Science* 18, no. 12 (2007): 1050–1057.

Forster, Edward Morgan. *Aspects of the Novel*. Boston: Houghton Mifflin Harcourt, 1985.

Foucault, Michel. *Confessions of the Flesh: The History of Sexuality*, vol. 4. New York: Pantheon, 2022.

Fredrickson, Barbara L. "Love: Positivity resonance as a fresh, evidence-based perspective on an age-old topic." *Handbook of Emotions* 4 (2016): 847–858.

Frevert, Ute. *The Politics of Humiliation: A Modern History*. Oxford: Oxford University Press, 2020.

Freytag, Gustav. *Die Technik des Dramas* [1886]. Stuttgart: Reclam, 1983.

Frijda, Nico H. "The psychologists' point of view." In *Handbook of Emotions*, edited by Michael Lewis, Jaennette M. Haviland-Jones, and Lisa Feldman Barrett, 68–87. New York: Guilford, 2008.

———. "The laws of emotion." *American Psychologist* 43, no. 5 (1988): 349.

Frijda, Nico H., Batja Mesquita, Joep Sonnemans, and Stephanie Van Goozen. "The duration of affective phenomena or emotions, sentiments and passions." *International Review of Studies on Emotion* 1 (1991): 187–225.

Friston, Karl. "The free-energy principle: A unified brain theory?" *Nature Reviews Neuroscience* 11, no. 2 (2010): 127–138.

Friston, Karl, James Kilner, and Lee Harrison. "A free energy principle for the brain." *Journal of Physiology-Paris* 100, no. 1–3 (2006): 70–87.

Friston, Karl, Francesco Rigoli, Dimitri Ognibene, Christoph Mathys, Thomas Fitzgerald, and Giovanni Pezzulo. "Active inference and epistemic value." *Cognitive Neuroscience* 6, no. 4 (2015): 187–214.

Fuchs, Thomas. *Ecology of the Brain: The Phenomenology and Biology of the Embodied Mind*. Oxford: Oxford University Press, 2018.

Gardner, Richard E., and Jonathan Daw. "Tulpamancy: A closeted community of imaginary-friend hobbyists." *Penn State McNair Journal* 42 (2017).

Garrido-Merchán, Eduardo C., José Luis Arroyo-Barrigüete, and Roberto Gozalo-Brihuela. "Simulating H. P. Lovecraft horror literature with the ChatGPT large language model." arXiv preprint arXiv:2305.03429 (2023).

Genette, Gérard. *Narrative Discourse: An Essay in Method.* Ithaca: Cornell University Press, 1980.

Gerrig, Richard J., and William G. Wenzel. "The role of inferences in narrative experiences." In *Inferences During Reading*, edited by Edward J. O'Brien, Ann E. Cook, and Robert F. Lorch, 362–385. Cambridge: Cambridge University Press, 2015.

Gess, Nicola, and Mireille Schnyder. "Staunen als Grenzphänomen: Eine Einführung." In *Staunen als Grenzphänomen*, edited by Nicola Gess, Mireille Schnyder, Hugues Marchal, and Johannes Bartuschat, 7–18. Paderborn: Wilhelm Fink Verlag, 2017.

Geulen, Eva. "Anagnorisis statt Identifikation (Raabes Altershausen)." In *Empathie und Erzählung*, edited by Claudia Breger and Fritz Breithaupt, 205–229. Freiburg: Rombach Verlag, 2010.

Gigerenzer, Gerd. *Gut Feelings: The Intelligence of the Unconscious.* London: Penguin, 2007.

Gigerenzer, Gerd, and Peter M. Todd. "Fast and frugal heuristics: The adaptive toolbox." In *Simple Heuristics That Make Us Smart*, edited by Gerd Gigerenzer and Peter M. Todd, 3–34. Oxford: Oxford University Press, 1999.

Gill, Michael J., and Phillip D. Getty. "On shifting the blame to humanity: Historicist narratives regarding transgressors evoke compassion for the transgressor but disdain for humanity." *British Journal of Social Psychology* 55, no. 4 (2016): 773–791.

Girard, René. *Violence and the Sacred*, translated by Patrick Gregory. Baltimore: Johns Hopkins University Press, 1977.

Glasser, Matthew F., Timothy S. Coalson, Emma C. Robinson, Carl D. Hacker, John Harwell, Essa Yacoub, Kamil Ugurbil et al. "A multi-modal parcellation of human cerebral cortex." *Nature* 536, no. 7615 (2016): 171–178.

Gleaves, David H., Steven M. Smith, Lisa D. Butler, and David Spiegel. "False and recovered memories in the laboratory and clinic: A review of experimental and clinical evidence." *Clinical Psychology: Science and Practice* 11, no. 1 (2004): 3–28.

Goethe, Johann Wolfgang von. *Wilhelm Meister's Apprenticeship and Travel*, translated by Thomas Carlyle. New York: A.L. Burt, 1840.

Goffman, Erving. *Stigma: Notes on the Management of Spoiled Identity.* Englewood Cliffs: Prentice-Hall, 1963.

———. *Presentations of Self in Everyday Life.* Garden City: Doubleday, 1959.

Gottschall, Jonathan. *The Story Paradox: How Our Love of Storytelling Builds Societies and Tears Them Down.* New York: Basic, 2021.

———. *The Storytelling Animal: How Stories Make Us Human.* Boston: Houghton Mifflin Harcourt, 2012.

Grätz, Manfred. *Das Märchen in der deutschen Aufklärung.* Stuttgart: J.B. Metzler, 1988.

Green, Melanie C. "Transportation into narrative worlds: The role of prior knowledge and perceived realism." *Discourse Processes* 38, no. 2 (2004): 247–266.

Green, Melanie C., Timothy C. Brock, and Geoff F. Kaufman. "Understanding media enjoyment: The role of transportation into narrative worlds." *Communication Theory* 14, no. 4 (2004): 311–327.

Green, Melanie C., Christopher Chatham, and Marc A. Sestir. "Emotion and transportation into fact and fiction." *Scientific Study of Literature* 2, no. 1 (2012): 37–59.

Griffiths, Thomas, Stephan Lewandowsky, and Michael L. Kalish. "The effects of cultural transmission are modulated by the amount of information transmitted." *Cognitive Science* 37, no. 5 (2013): 953–967.

Gumbrecht, Hans Ulrich. *Production of Presence: What Meaning Cannot Convey.* Palo Alto: Stanford University Press, 2004.

Hahamy, Avital, Haim Dubossarsky, and Timothy E. J. Behrens. "The human brain reactivates context-specific past information at event boundaries of naturalistic experiences." *Nature Neuroscience* (2023): 1–10.

Haidt, Jonathan. *The Righteous Mind: Why Good People Are Divided by Politics and Religion.* New York: Vintage, 2012.

Hamilton, Andrew, and Fritz Breithaupt. "These things called event: Toward a unified narrative theory of events." *Language and Computing* 37 (2013): 65–87.

Hanich, Julian, Valentin Wagner, Mira Shah, Thomas Jacobsen, and Winfried Menninghaus. "Why we like to watch sad films: The pleasure of being moved in aesthetic experiences." *Psychology of Aesthetics, Creativity, and the Arts* 8, no. 2 (2014): 130.

Harari, Yuval Noah. *Sapiens: A Brief History of Humankind.* New York: HarperCollins, 2015.

He, Tianyou, Fritz Breithaupt, Sandra Kübler, and Thomas T. Hills. "Quantifying the retention of emotions across story retellings." *Scientific Reports* 13, no. 1 (2023): 2448.

Heath, Chip, Chris Bell, and Emily Sternberg. "Emotional selection in memes: The case of urban legends." *Journal of Personality and Social Psychology* 81, no. 6 (2001): 1028.

Heidegger, Martin. *Plato: Sophistes: Marburg Lecture, Winter Semester 1924/1925*. Frankfurt: Klostermann, 2018.

Hendriks-Jansen, Horst. *Catching Ourselves in the Act: Situated Activity, Interactive Emergence, Evolution, and Human Thought*. Cambridge: MIT Press, 1996.

Hiskes, Benjamin, Milo Hicks, Samuel Evola, Cameron Kincaid, and Fritz Breithaupt. "Multiversionality: Considering multiple possibilities in the processing of a narrative." *Review of Philosophy and Psychology* 14 (2022): 1099–1224.

Hoffmann, Dirk, Christopher D. Standish, Marcos García-Diez, Paul B. Pettitt, James A. Milton, João Zilhão, Javier J. Alcolea-González et al. "U-Th dating of carbonate crusts reveals Neandertal origin of Iberian cave art." *Science* 359, no. 6378 (2018): 912–915.

Hofstadter, Douglas. "Gödel, Escher, Bach, and AI." *The Atlantic* (November 15, 2023), https://www.theatlantic.com/ideas/archive/2023/07/godel-escher -bach-geb-ai/674589/.

———. *I Am a Strange Loop*. New York: Basic, 2007.

———. *Gödel, Escher, Bach*. London: Harvester, 1979.

Hohwy, Jakob. *The Predictive Mind*. Oxford: Oxford University Press, 2013.

Homer. *The Odyssey*, translated by Samuel Butler. London: A.C. Fifield, 1900.

Hosseini, Arian, Siva Reddy, Dzmitry Bahdanau, R. Devon Hjelm, Alessandro Sordoni, and Aaron Courville. "Understanding by understanding not: Modeling negation in language models." arXiv preprint arXiv:2105.03519 (2021).

Hunt, Lynn. *Inventing Human Rights*. New York: Norton, 2007.

Hutchinson, Benjamin J., and Lisa Feldman Barrett. "The power of predictions: An emerging paradigm for psychological research." *Current Directions in Psychological Science* 28, no. 3 (2019): 280–291.

Hwang, Hyisung C., David Matsumoto, Hiroshi Yamada, Aleksandra Kostić, and Juliana V. Granskaya. "Self-reported expression and experience of triumph across four countries." *Motivation and Emotion* 40, no. 5 (2016): 731–739.

Illouz, Eva. *The End of Love: A Sociology of Negative Relations*. Oxford: Oxford University Press, 2019.

Iser, Wolfgang. *The Act of Reading: A Theory of Aesthetic Response*. London: Routledge, 1978.

Jagiello, Robert D., and Thomas T. Hills. "Bad news has wings: Dread risk mediates social amplification in risk communication." *Risk Analysis* 38, no. 10 (2018): 2193–2207.

Jannidis, Fotis. *Figur und Person: Beitrag zu einer historischen Narratologie.* Berlin: De Gruyter, 2004.

Jolles, André. *Simple Forms: Legend, Saga, Myth, Riddle, Saying, Case, Memorabile, Fairytale, Joke* [1930]. London: Verso, 2017.

Jung-Stilling, Johann Heinrich. *Henrich Stillings Jugend, Jünglingsjahre, Wanderschaft und häusliches Leben.* Stuttgart: Reclam, 1997.

Kaland, Nils, Annette Møller-Nielsen, Lars Smith, Erik Lykke Mortensen, Kirsten Callesen, and Dorte Gottlieb. "The strange stories test: A replication study of children and adolescents with Asperger syndrome." *European Child and Adolescent Psychiatry* 14 (2005): 73–82.

Kanske, Philipp. "The social mind: Disentangling affective and cognitive routes to understanding others." *Interdisciplinary Science Reviews* 43, no. 2 (2018): 115–124.

Kashima, Yoshihisa. "Maintaining cultural stereotypes in the serial reproduction of narratives." *Personality and Social Psychology Bulletin* 26, no. 5 (2000): 594–604.

Kashima, Yoshihisa, Anthony Lyons, and Anna Clark. "The maintenance of cultural stereotypes in the conversational retelling of narratives." *Asian Journal of Social Psychology* 16, no. 1 (2013): 60–70.

Keen, Suzanne. "A theory of narrative empathy." *Narrative* 14, no. 3 (2006): 207–236.

Kelley, Harold H., John G. Holmes, Norbert L. Kerr, Harry T. Reis, Caryl E. Rusbult, and Paul Van Lange. *An Atlas of Interpersonal Situations.* Cambridge: Cambridge University Press, 2003.

Kelley, Karol. "A modern Cinderella." *Journal of American Culture* 17, no. 1 (1994): 87.

Kensinger, Elizabeth A., and Daniel L. Schacter. "Memory and emotion." *Handbook of Emotions* 4 (2016): 564–578.

Kermode, Frank. *The Sense of an Ending: Studies in the Theory of Fiction with a New Epilogue.* Oxford: Oxford University Press, 2000.

Kidd, Celeste, and Benjamin Y. Hayden. "The psychology and neuroscience of curiosity." *Neuron* 88, no. 3 (2015): 449–460.

Kiss, Orsolya. "Reinventing the plot: J. C. Gottsched's *Dying Cato.*" *Deutsche Vierteljahrsschrift für Literaturwissenschaft und Geistesgeschichte* 84, no. 4 (2010): 507–525.

Knill, David C., and Alexandre Pouget. "The Bayesian brain: The role of uncertainty in neural coding and computation." *Trends in Neurosciences* 27, no. 12 (2004): 712–719.

Knutson, Brian, and Jeffrey C. Cooper. "The lure of the unknown." *Neuron* 51, no. 3 (2006): 280–282.

Kobayashi, Hiromi, and Shiro Kohshima. "Unique morphology of the human eye and its adaptive meaning: Comparative studies on external morphology of the primate eye." *Journal of Human Evolution* 40, no. 5 (2001): 419–435.

Kolmar, Martin. *Grenzbeschreibungen: Vom Sinn, dem gelingenden Leben und unserem Umgang mit Natur.* Cologne: Böhlau, 2021.

Koschorke, Albrecht. *Fact and Fiction: Elements of a General Theory of Narrative.* Berlin: De Gruyter, 2018.

———. *Körperströme und Schriftverkehr: Mediologie des 18. Jahrhunderts.* Munich: Fink, 1999.

Kosinski, Michal. "Theory of mind may have spontaneously emerged in large language models." arXiv preprint arXiv:2302.02083 (2023).

Krawietz, Sabine A., and Andrea K. Tamplin. "Walking through doorways causes forgetting: Further explorations." *Quarterly Journal of Experimental Psychology* 64, no. 8 (2011): 1632–1645.

Krupp, Anthony. *Reason's Children: Childhood in Early Modern Philosophy.* Lewisburg, Pa.: Bucknell University Press, 2009.

Kucyi, A., N. Anderson, T. Bounyarith, D. Braun, L. Shareef-Trudeau, I. Treves, . . . and S. M. Hung. "Individual variability in neural representations of mind-wandering." BioRxiv: The Preprint Server for Biology (posted January 22, 2024).

Kukkonen, Karin. *Probability Designs: Literature and Predictive Processing.* Oxford: Oxford University Press, 2020.

Kummer, Hans. *Social Organization of Hamadyras Baboons.* Chicago: University of Chicago Press, 1968.

Lagrange, Victoria. "Individualized communal experience: Players of *Detroit: Become Human.*" November 1, 2023, http://digra.org:9998/DiGRA_2023_CR_9908.pdf.

Lagrange, Victoria, Denizhan Pak, Vanessa Denny, Cameron Kincaid, Sara Konrath, and Fritz Breithaupt. "Pawn-playing and biased empathy: Interactive fiction promotes single-perspective empathy, linear fiction multi-perspective empathy." (2024), forthcoming.

Laland, Kevin N., and Kim Sterelny. "Perspective: Seven reasons (not) to neglect niche construction." *Evolution* 60, no. 9 (2006): 1751–1762.

Lauer, Gerhard. "Das Erdbeben von Lissabon: Ereignis, Wahrnehmnung und Deutung im Zeitalter der Aufklärung." In *Das Erdbeben von Lissabon und der Katastrophendiskurs im 18. Jahrhundert*, edited by Gerhard Lauer and Thorsten Unger, 223–236. Göttingen: Wallstein, 2008.

Lazer, David, Matthew A. Baum, Yochai Benkler, Adam J. Berinsky, Kelly M. Greenhill, Filippo Menczer, Miriam J. Metzger et al. "The science of fake news." *Science* 359, no. 6380 (2018): 1094–1096.

Leite, Abe, and Eduardo J. Izquierdo. "Generating reward structures on a parameterized distribution of dynamics tasks." *ALIFE 2021: The 2021 Conference on Artificial Life*. Cambridge: MIT Press, 2021.

Lévinas, Emmanuel. "The trace of the other." In *Deconstruction in Context: Literature and Philosophy*, edited by Mark C. Taylor, 345–359. Chicago: University of Chicago Press, 1986.

Lewis, Michael. "Self-conscious emotions: Embarrassment, pride, shame, and guilt." *Handbook of Emotions* 4 (2016): 792–814.

Libet, Benjamin. *Mind Time: The Temporal Factor in Consciousness*. Cambridge: Harvard University Press, 2009.

Litman, Jordan A. "Curiosity and metacognition." In *Metacognition: New Research Developments*, edited by Clayton B. Larson, 105–116. Hauppauge, N.Y.: Nova Science, 2009.

Lorenz, Konrad. *Er redete mit dem Vieh, den Vögeln und den Fischen*. Munich: dtv, 1964: 84–95.

Luhmann, Niklas. *Introduction to Systems Theory*, translated by Peter Gilgen. Cambridge: Polity, 2013.

Luhrmann, Tanya M., Ben Alderson-Day, Ann Chen, Philip Corlett, Quinton Deeley, David Dupuis, Michael Lifshitz et al. "Learning to discern the voices of gods, spirits, tulpas, and the dead." *Schizophrenia Bulletin* 49, Supplement 1 (2023): S3–S12.

Madeira, Jody Lyneé. *Killing McVeigh: The Death Penalty and the Myth of Closure*. New York: NYU Press, 2012.

Magliano, Joseph P., and Jeffrey M. Zacks. "The impact of continuity editing in narrative film on event segmentation." *Cognitive Science* 35, no. 8 (2011): 1489–1517.

Mahr, Johannes B., and Gergely Csibra. "Why do we remember? The communicative function of episodic memory." *Behavioral and Brain Sciences* 41 (2018): 1–65.

Mandler, Jean M., and Nancy S. Johnson. "Remembrance of things parsed: Story structure and recall." *Cognitive Psychology* 9, no. 1 (1977): 111–151.

Margulis, Elizabeth Hellmuth. *On Repeat: How Music Plays the Mind*. Oxford: Oxford University Press, 2014.

Martínez, Angeles M. "Storyworld possible selves and the phenomenon of narrative immersion: Testing a new theoretical construct." *Narrative* 22, no. 1 (2014): 110–131.

Martínez, Matías. "Figuration." In *Handbuch Erzählliteratur: Theorie, Analyse, Geschichte*, edited by Matías Martínez, 145–150. Wiesbaden: Springer-Verlag, 2011.

Masters, John. *Casanova*. London: Joseph, 1969.

Matuschek, Stefan. *Über das Staunen: Eine ideengeschichtliche Analyse*. Tübingen: Niemeyer, 1991.

McAdams, Dan P. "'First we invented stories, then they changed us': The evolution of narrative identity." *Evolutionary Studies in Imaginative Culture* 3, no. 1 (2019): 1–18.

McAdams, Dan P., and Kate McLean. "Narrative identity." In *Handbook of Identity Theory and Research*, 99–115. New York: Springer, 2011.

McMillan, Rebecca, Scott Barry Kaufman, and Jerome L. Singer. "Ode to positive constructive daydreaming." *Frontiers in Psychology* 4 (2013): 626.

Mellmann, Katja. "Evolutionary proto-forms of literary behavior." *Ethology and the Arts* (2013): 389–404.

Menninghaus, Winfried, Valentin Wagner, Julian Hanich, Eugen Wassiliwizky, Milena Kuehnast, and Thomas Jacobsen. "Towards a psychological construct of being moved." In *PloS One* 10, no. 6 (2015): e0128451.

Merrill, Julia, Anna Czepiel, Lea T. Fink, Jutta Toelle, and Melanie Wald-Fuhrmann. "The aesthetic experience of live concerts: Self-reports and psychophysiology." *Psychology of Aesthetics, Creativity, and the Arts* 17, no. 2 (2023): 134.

Mesoudi, Alex. *Cultural Evolution: How Darwinian Theory Can Explain Human Culture and Synthesize the Social Sciences*. Chicago: University of Chicago Press, 2011.

Mesoudi, Alex, and Andrew Whiten. "The multiple roles of cultural transmission experiments in understanding human cultural evolution." *Philosophical Transactions of the Royal Society B: Biological Sciences* 363, no. 1509 (2008): 3489–3501.

Mesoudi, Alex, Andrew Whiten, and Robin Dunbar. "A bias for social information in human cultural transmission." *British Journal of Psychology* 97, no. 3 (2006): 405–423.

Miall, David. "Anticipation and feeling in literary response: A neuropsychological perspective." *Poetics* 23, no. 4 (1995): 275–298.

Miller, Geoffrey. *The Mating Mind: How Sexual Choice Shaped the Evolution of Human Nature*. New York: Anchor, 2011.

Möller, Reinhard M. "Ästhetiken gegenstandsbezogenen Staunens im 18. Jahrhundert." In *Staunen als Grenzphänomen*, edited by Nicola Gess, Mireille Schnyder, Hugues Marchal, and Johannes Bartuschat, 107–124. Paderborn: Wilhelm Fink Verlag, 2017.

Monroy, Claire D., Sarah A. Gerson, and Sabine Hunnius. "Translating visual information into action predictions: Statistical learning in action and nonaction contexts." *Memory and Cognition* 46, no. 4 (2018): 600–613.

Mooneyham, Benjamin W., and Jonathan W. Schooler. "The costs and benefits of mind-wandering: A review." *Canadian Journal of Experimental Psychology/Revue canadienne de psychologie expérimentale* 67, no. 1 (2013): 11–18.

Moretti, Franco. *Signs Taken for Wonders: Essays in the Sociology of Literary Forms.* London: Verso, 1983.

Mounk, Yascha. *The Identity Trap: A Story of Ideas and Power in Our Time.* New York: Random House, 2023.

Moussaïd, Mehdi, Henry Brighton, and Wolfgang Gaissmaier. "The amplification of risk in experimental diffusion chains." *Proceedings of the National Academy of Sciences* 112, no. 18 (2015): 5631–5636.

Musil, Robert. *A Man Without Qualities,* translated by Sophie Wilkins and Burton Pike. New York: Vintage, 1995.

Nabi, Robin. "The case for emphasizing discrete emotions in communication research." *Communication Monographs* 77 (2010): 153–159.

Nagel, Thomas. "What is it like to be a bat?" *Philosophical Review* 83, no. 4 (1974): 435–450.

Ngai, Sianne. *Ugly Feelings.* Cambridge: Harvard University Press, 2005.

Nicholas, Franklin T., Kenneth A. Norman, Charan Ranganath, Jeffrey M. Zacks, and Samuel J. Gershman. "Structured event memory: A neuro-symbolic model of event cognition." *Psychological Review* 127, no. 3 (2020): 327.

Noordewier, Marret K., Sascha Topolinski, and Eric Van Dijk. "The temporal dynamics of surprise." *Social and Personality Psychology Compass* 10, no. 3 (2016): 136–149.

Norberg, Jakob. *The Brothers Grimm and the Making of German Nationalism.* Cambridge: Cambridge University Press, 2022.

Norenzayan, Ara, Scott Atran, Jason Faulkner, and Mark Schaller. "Memory and mystery: The cultural selection of minimally counterintuitive narratives." *Cognitive Science* 30, no. 3 (2006): 531–553.

Nussbaum, Martha C. *Political Emotions: Why Love Matters for Justice.* Cambridge: Harvard University Press, 2013.

Nyhof, Melanie, and Justin Barrett. "Spreading non-natural concepts: The role of intuitive conceptual structures in memory and transmission of cultural materials." *Journal of Cognition and Culture* 1, no. 1 (2001): 69–100.

Opfer, John E., and Susan A. Gelman. "Development of the animate-inanimate distinction." *Wiley-Blackwell Handbook of Childhood Cognitive Development,* 213–238. Malden, Mass.: Wiley-Blackwell, 2011.

Ortony, Andrew, Gerald L. Clore, and Allan Collins. *The Cognitive Structure of Emotions*. Cambridge: Cambridge University Press, 1990.

Perner, Josef, Michael Huemer, and Brian Leahy. "Mental files and belief: A cognitive theory of how children represent belief and its intensionality." *Cognition* 145 (2015): 77–88.

Perrault, Charles. *The Complete Fairy Tales*, translated by Christopher Betts. New York: Oxford University Press, 2009.

Pestalozzi, Johann Heinrich. *Tagebuch Pestalozzis über die Erziehung seines Sohnes, 7–18*. Munich: Winkler, 1977.

Pettijohn, Kyle A., and Gabriel A. Radvansky. "Narrative event boundaries, reading times, and expectation." *Memory and Cognition* 44, no. 7 (2016): 1064–1075.

Pinker, Steven. *The Better Angels of Our Nature: Why Violence Has Declined*. New York: Viking, 2011.

———. *How the Mind Works*. New York: Norton, 1996.

Pistrol, Florian. "Vulnerability: Erläuterungen zu einem Schlüsselbegriff im Denken Judith Butlers." *Journal of Practical Philosophy* 3, no. 1 (2016): 233–272.

Princefirestorm. "Switching and co-fronting." November 11, 2023, https://www.wattpad.com/829536502-tulpamancy-switching-and-co-fronting.

Radvansky, Gabriel A., and Jeffrey M. Zacks. "Event boundaries in memory and cognition." *Current Opinion in Behavioral Sciences* 17 (2017): 133–140.

———. *Event Cognition*. New York: Oxford University Press, 2014.

Raichle, Marcus E., Ann Mary MacLeod, Abraham Z. Snyder, William J. Powers, Debra A. Gusnard, and Gordon L. Shulman. "A default mode of brain function." *Proceedings of the National Academy of Sciences* 98, no. 2 (2001): 676–682.

Ranganath, Charan. *Why We Remember: Unlocking Memory's Power to Hold on to What Matters*. New York: Doubleday, 2024.

Reber, Rolf. *Critical Feeling*. Cambridge: Cambridge University Press, 2016.

Reisenzein, Rainer. "Exploring the strength of association between the components of emotion syndromes: The case of surprise." *Cognition and Emotion* 14, no. 1 (2000): 1–38.

Rigney, Ann. "The point of stories: On narrative communication and its cognitive functions." *Poetics Today* 13, no. 2 (1992): 263–283.

Rimé, Bernard. "Emotion elicits the social sharing of emotion: Theory and empirical review." *Emotion Review* 1 (2009): 60–85.

Rogers, Benjamin A., Herrison Chicas, John Michael Kelly, Emily Kubin, Michael S. Christian, Frank J. Kachanoff, Jonah Berger, Curtis Puryear, Dan P. McAdams, and Kurt Gray. "Seeing your life story as a Hero's Journey increases meaning in life." *Journal of Personality and Social Psychology* 125, no. 4 (2023): 752–778.

Rosa, Hartmut. *Social Acceleration: A New Theory of Modernity.* New York: Columbia University Press, 2013.

Roseman, Ira J., Martin S. Spindel, and Paul E. Jose. "Appraisals of emotion-eliciting events: Testing a theory of discrete emotions." *Journal of Personality and Social Psychology* 59, no. 5 (1990): 899–915.

Roskies, Adina L. "How does neuroscience affect our conception of volition?" *Annual Review of Neuroscience* 33 (2010): 109–130.

Rothstein, Bret L. *The Shape of Difficulty: A Fan Letter to Unruly Objects.* University Park: Penn State University Press, 2019.

Rumelhart, David. "Understanding and summarizing brief stories." In *Basic Processes in Reading: Perception and Comprehension,* edited by D. La Berge and S. J. Samuels, 265–302. Hillsdale, N.J.: Lawrence Erlbaum Associates, 1977.

Rüth, Axel. "Representing wonder in medieval miracle narratives." *Modern Language Notes* 126, no. 4 (2011): 89–114.

Sargent, Jesse Q., Jeffrey M. Zacks, David Z. Hambrick, Rose T. Zacks, Christopher A. Kurby, Heather R. Bailey, Michelle L. Eisenberg, and Taylor M. Beck. "Event segmentation ability uniquely predicts event memory." *Cognition* 129, no. 2 (2013): 241–255.

Scarry, Elaine. *Dreaming by the Book.* Princeton: Princeton University Press, 2001.

Schapp, Wilhelm. *In Geschichten verstrickt: Zum Sein von Ding und Mensch.* Hamburg: Meiners, 1953.

Schechner, Richard. *Between Theater and Anthropology* [1985]. Philadelphia: University of Pennsylvania Press, 2010.

Schechter, Elizabeth. "The unity of consciousness." *Routledge Handbook of Consciousness,* 366–378. London: Routledge, 2018.

Scherer, Klaus R., Angela Schorr, and Tom Johnstone. *Appraisal Processes in Emotion: Theory, Methods, Research.* Oxford: Oxford University Press, 2001.

Schilbach, Leonhard. "On the relationship of online and offline social cognition." *Frontiers in Human Neuroscience* 8 (2014): 278.

Schindler, Ines, Georg Hosoya, Winfried Menninghaus, Ursula Beermann, Valentin Wagner, Michael Eid, and Klaus R. Scherer. "Measuring aesthetic emotions: A review of the literature and a new assessment tool." *PloS One* 12, no. 6 (2017): e0178899.

Schmid, Wolf. "Eventfulness and repetitiveness: Two aesthetics of storytelling." In *Emerging Vectors of Narratology,* edited by Per Krogh Hansen, John Pier, Philippe Roussin, and Wolf Schmid, 229–246. Berlin: De Gruyter, 2017.

————. *Mentale Ereignisse: Bewusstseinsveränderungen in europäischen Erzähl-werken vom Mittelalter bis zur Moderne.* Berlin: De Gruyter, 2017.

Schulz, Armin, and Gert Hübner. "Mittelalter." In *Handbuch Erzählliteratur: Theorie, Analyse, Geschichte,* edited by Matías Martínez, 184–204. Wiesbaden: Springer-Verlag, 2011.

Schulze, Hagen. *Germany: A New History.* Cambridge: Harvard University Press, 1998.

Sen, Amartya. *The Idea of Justice.* Cambridge: Harvard University Press, 2009.

Serafini, Randal A., Kerri D. Pryce, and Venetia Zachariou. "The mesolimbic dopamine system in chronic pain and associated affective comorbidities." *Biological Psychiatry* 87, no. 1 (2020): 64–73.

Sheehan, Jonathan. *The Enlightenment Bible: Translation, Scholarship, Culture.* Princeton: Princeton University Press, 2005.

Shiller, Robert. *Narrative Economics: How Stories Go Viral and Drive Major Economic Events.* Princeton: Princeton University Press, 2020.

Slocombe, Katie E., and Klaus Zuberbühler. "Functionally referential communication in a chimpanzee." *Current Biology* 15, no. 19 (2005): 1779–1784.

Smallwood, Jonathan, and Jessica Andrews-Hanna. "Not all minds that wander are lost: The importance of a balanced perspective on the mind-wandering state." *Frontiers in Psychology* 4 (2013): 441.

Smuts, Aaron. "The paradox of suspense." *Stanford Encyclopedia of Philosophy,* 2009; https://plato.stanford.edu/archives/fall2009/entries/paradox-suspense/.

Soni, Vivasvan. "Trials and tragedies: The literature of unhappiness (a model for reading narratives of suffering)." *Comparative Literature* 59, no. 2 (2007): 119–139.

Speer, Nicole K., Jeffrey M. Zacks, and Jeremy R. Reynolds. "Human brain activity time-locked to narrative event boundaries." *Psychological Science* 18, no 5 (2007): 449–455.

Starr, G. Gabrielle. "Theorizing imagery, aesthetics, and doubly directed states." In *The Oxford Handbook of Cognitive Literary Studies,* edited by Lisa Zunshine, 246–268. Oxford: Oxford University Press, 2015.

Stawarczyk, David, Matthew A. Bezdek, and Jeffrey M. Zacks. "Event representations and predictive processing: The role of the midline default network core." *Topics in Cognitive Science* 13, no. 1 (2021): 164–186.

Steedman, Carolyn. *Strange Dislocations: Childhood and the Idea of Human Interiority, 1780–1930.* Cambridge: Harvard University Press, 1995.

Strawson, Galen. "Against narrativity." *Ratio* 17, no. 4 (2004): 428–452.

Stubbersfield, Joseph M., Jamshid J. Tehrani, and Emma G. Flynn. "Chicken tumors and a fishy revenge: Evidence for emotional content bias in the cumulative recall of urban legends." *Journal of Cognition and Culture* 17, no. 1–2 (2017): 12–26.

———."Serial killers, spiders and cybersex: Social and survival information bias in the transmission of urban legends." *British Journal of Psychology* 106, no. 2 (2015): 288–307.

Sütterlin, Nicole. *Poetik der Wunde: Zur Entdeckung des Traumas in der Literatur der Romantik.* Göttingen: Wallstein Verlag, 2019.

Szondi, Peter. *An Essay on the Tragic,* translated by Paul Fleming. Palo Alto: Stanford University Press, 2002.

Tatar, Maria. *The Hard Facts of the Grimms' Fairy Tales: Expanded Edition.* Princeton: Princeton University Press, 2019.

Tobin, Vera. *Elements of Surprise: Our Mental Limits and the Satisfactions of Plot.* Cambridge: Harvard University Press, 2018.

Todd, Peter M., and Gerd Gigerenzer. "Précis of simple heuristics that make us smart." *Behavioral and Brain Sciences* 23, no. 5 (2000): 727–741.

Tomasello, Michael. *Origins of Human Communication.* Cambridge: MIT Press, 2010.

———. *Why We Cooperate.* Cambridge: MIT Press, 2009.

Tooby, John, and Leda Cosmides. "Does beauty build adapted minds? Toward an evolutionary theory of aesthetics, fiction, and the arts." *SubStance* 30, no. 1 (2001): 6–27.

Topolinski, Sascha, and Rolf Reber. "Gaining insight into the 'Aha' experience." *Current Directions in Psychological Science* 19, no. 6 (2010): 402–405.

Topolinski, Sascha, and Fritz Strack. "Corrugator activity confirms immediate negative affect in surprise." *Frontiers in Psychology* 6 (2015): 134.

Tulving, Endel. "Episodic memory: From mind to brain." *Annual Review of Psychology* 53, no. 1 (2002): 1–25.

———. "Episodic and semantic memory." In *Organization of Memory,* edited by Endel Tulving and W. Donaldson, 381–403. New York: Academic, 1972.

Türk, Johannes. *Zur Immunität der Literatur.* Frankfurt: Fischer, 2011.

Turner, Mark. *The Literary Mind: The Origins of Thought and Language.* Oxford: Oxford University Press, 1996.

Vago, David R., and Fadel Zeidan. "The brain on silent: Mind wandering, mindful awareness, and states of mental tranquility." *Annals of the New York Academy of Sciences* 1373, no. 1 (2016): 96.

Varnum, Michael, and Igor Grossmann. "Cultural change: The how and the why." *Perspectives on Psychological Science* 12, no. 6 (2017): 956–972.

Vermeule, Blakey. *Why Do We Care About Literary Characters?* Baltimore: Johns Hopkins University Press, 2010.

Voss, Christiane. *Narrative Emotionen*. Berlin: De Gruyter, 2004.

Wagoner, Brady. *The Constructive Mind: Bartlett's Psychology in Reconstruction*. Cambridge: Cambridge University Press, 2017.

Warburton, Edward C. "Of meanings and movements: Re-languaging embodiment in dance phenomenology and cognition." *Dance Research Journal* 43, no. 2 (2011): 65–84.

Wegner, Daniel M., and Kurt Gray. *The Mind Club: Who Thinks, What Feels, and Why It Matters*. London: Penguin, 2017.

Winograd, Eugene, and Ulric Neisser. *Affect and Accuracy in Recall: Studies of 'Flashbulb' Memories*. Cambridge: Cambridge University Press, 2006.

Wolf, Maryanne. *Proust and the Squid: The Story and Science of the Reading Brain*. New York: Harper Perennial, 2008.

Wood, Samantha, and Justin Wood. "One-shot object parsing in newborn chicks." *Journal of Experimental Psychology: General* 150, no. 11 (2021): 2408.

Woodward, Claire, Ben Hiskes, and Fritz Breithaupt. "Spontaneous side-taking drives memory, empathy, and author attribution in conflict narratives." *Discover Psychology* 4, no. 1 (2024): 1–13.

Yang, Xianjun, Yan Li, Xinlu Zhang, Haifeng Chen, and Wei Cheng. "Exploring the limits of ChatGPT for query or aspect-based text summarization." arXiv preprint arXiv:2302.08081 (2023).

Yon, Daniel, Cecilia Heyes, and Clare Press. "Beliefs and desires in the predictive brain." *Nature Communications* 11, no. 1 (2020): 1–4.

Zacks, Jeffrey M., Nicole K. Speer, Khena M. Swallow, Todd S. Braver, and Jeremy R. Reynolds. "Event perception: A mind-brain perspective." *Psychological Bulletin* 133, no. 2 (2007): 273.

Zacks, Jeffrey M., Barbara Tversky, and Gowri Iyer. "Perceiving, remembering, and communicating structure in events." *Journal of Experimental Psychology: General* 130, no. 1 (2001): 29.

Zanger, Anat. *Film Remakes as Ritual and Disguise*. Amsterdam: Amsterdam University Press, 2006.

Zipes, Jack. *Fairy Tales and the Art of Subversion*. London: Routledge, 2012.

Žižek, Slavoj. *Event: A Philosophical Journey Through a Concept*. New York: Melville House, 2014.

Zunshine, Lisa. *Why We Read Fiction: Theory of Mind and the Novel*. Columbus: Ohio State University Press, 2006.

Zwaan, Rolf A., Mark C. Langston, and Arthur C. Graesser. "The construction of situation models in narrative comprehension: An event-indexing model." *Psychological Science* 6, no. 5 (1995): 292–297.

INDEX

44, 202, 204, 208, 218, 219; and
empathy, 203, 219; evolution of,
205–206; and narrative thinking, 206, 216
Cook, Amy: on casting, 174
counter-narratives. *See* multiversionality
Covid-19 pandemic: narrative
patterns associated with,
161–162; narratives relating
to, 159–163
crises: collective narratives of,
146–148, 150, 163–164; narrative thinking as response to,
146–148; September 11th as,
148–151
Csibra, Gergely, 252n28
cultural appropriation, 57
cultural evolution: of narratives,
19–21; and serial reproduction,
52
cultural misunderstandings,
135–136
culture: concept of, 20–21

Damasio, Antonio: *Descartes' Error*,
241–242n55
Darwin, Charles, 19
Daston, Lorraine, 119
Dawkins, Richard, 19
daydreams: emotions associated
with, 243n13; with narrative
structure, 113–115
Dennett, Daniel, 178, 205
depression, people with: and cognitive distortions, 181–182
Detroit: Become Human, 169–170,
198
deus ex machina, 39
disgust, stories of, 88, 89

dopamine receptors: and emo-
tional rewards, 109
Dorsch, Fabian, 113–114
Dotan, Shiman: *The Settlers*, 228
Dubourg, Edgar, 133
Dumas, Alexandre: *The Count of
Monte Cristo*, 35
Durkheim, Émile, 147; theory of
culture formulated by, 50

education: and the mutability of
humans, 62–63
embarrassment. *See* shame and
embarrassment
emotional appraisals: as aspect
of narrative, 16, 43, 78–80, 86,
97–98, 102–104
emotional rewards, 12; and human
behavior, 109; and narrative
thinking, 107–113, 142–145
emotions, 3–4; appraisal theory of,
43; and causality, 78; as central
in narratives, 78–90, 101–105;
as contagious, 217; as reward,
12; and sense of community,
10–11; and serial reproduction,
78–79, 80–81, 101–105
empathy: collective, 203, 231; and
mobility of consciousness, 203,
219; redirecting, 250–251n22.
See also co-experience
Ems Dispatch: impact of, 47–48
episodicity, 30
episodic memories, 29, 252n28;
and semantic knowledge, 30–31
eroticism: as narrative emotion,
138–141
events: and narrative, 131–132; of
observation, 129; segmentation,
31–37

evolutionary theory: as applied to narrative thinking, 19–21

expectations: as active making of predictions, 193–196; of emotional rewards, 12; as traps, 4–6

experience, 1–3, 28–29, 42. *See also* co-experience

Experimental Humanities Laboratory, 10

fairy tales. *See* myths and fairy tales

false memory syndrome, 104

family roles: as traps, 5

fear: as narrative emotion, 144

Feldman Barrett, Lisa, 188, 190

fiction, interactive, 169, 198, 248n7

Flaubert, Gustave, 155

Flesch, William, 121, 123

Floyd, George: differing perspectives on, 226

Forster, E. M.: *Aspects of the Novel*, 237n32

Foucault, Michel, 139

Freud, Sigmund, 65, 114; *Beyond the Pleasure Principle*, 156; on characters in our mind, 172–173; *Totem and Taboo*, 50

Freytag, Gustav: on the middle of the narrative arc, 39–45, 123

Freytag's pyramid, 40–44, 73, 84, 85

Frijda, Nico: on emotion, 43

Game of Thrones, 9, 120–121

Gandhi, Mahatma, 115

Genette, Gérard, 155, 222, 224–225

Geulen, Eva, 130

global financial crisis (2008): villains of, 158–159

Goethe, Johann Wolfgang von, 59, 158; *Wilhelm Meister's Apprenticeship and Travel*, 132, 133, 178

Google Glass, 103

Gottschall, Jonathan, 227

Gottsched, Johann Christoph: *The Dying Cato*, 75

Gray, Kurt: *The Mind Club*, 249n23

Grimm Brothers' fairy tales, 50, 58, 59–60, 78, 105; danger and threat as aspects of, 68–69; distinguishing aspects of their narratives, 60 62; retelling of, 101–102; test narrative for characters in, 71–72; vulnerability and cunning of characters in, 69–74, 77

Grimmelshausen, Hans Jacob Christoph von: *The Adventures of Simplicius Simplicissimus*, 60–62; *Trutz Simplex*, 60–62

Haidt, Jonathan, 122

"Hansel and Gretel," 59, 72, 73

Harry Potter books, 65

Heidegger, Martin, 215

Hemingway, Ernest, 229

Herder, Johann Gottfried, 50

hindsight knowledge, 50

Hoffmann, E. T. A., 65

Hofstadter, Douglas, 100, 205

Holmes, Elizabeth, 154–155

Homer: *Odyssey*, 71

hope, 109, 193–194

Houellebecq, Michel, 155

How to Marry a Millionaire (film), 141

Qian, Sima, 178

Ranganath, Charan, 104
rationalization, and rationality: as
applied to narrative memory,
54–58, 82
Reber, Rolf, 247n71
reception and communication, 17
receptivity, cultivation of, 211–214
recognition, and revelation (anagnorisis): experience of, 127–131
repetition compulsion: following
trauma, 149–150
retelling of stories: and archetypes,
50; and changes from one to the
next, 101; by ChatGPT, 90–100;
culture as factor in, 50; emotion
as factor in, 78–90, 97, 101–106;
phases of, 101–105; processes
involved in, 48–51; and the
purpose of a story, 49–50; as
schema formation, 50–51, 52,
104; and scientific inquiry,
48–51; surprise as factor in,
89–90, 105
rewards, emotions as, 107–109,
142–143
Richardson, Samuel: *Clarissa*, 169
risk assessment, stories of, 88–89
ritualization, 50
Robin Hood, 71
Rogers, Benjamin A., 153
Romeo and Juliet (Shakespeare), 125
Rooney, Sally, 155
Rousseau, Jean-Jacques, 63
Rowling, J. K., 133
Russian fairy tales, 17

Sacks, Oliver, 178
Salzmann, Christian Gotthilf, 63

Schechner, Richard, 211
schema: as applied to Bartlett's
theories, 52; formation, and
retelling, 50–51; as used in
psychology, 18
Schiller, Friedrich: *Wallenstein*, 40
Schmid, Wolf, 132, 222–223
scripts: and expectations, 18–19
segmentation: and memory,
32–33, 46; of temporal processes, 27–29, 31, 36–37, 46
self-image: as a trap, 6, 179–183
self-representations (as narrative),
177–179
semantic knowledge, 29; and
episodic memories, 30–31
Sen, Amartya, 124
sentiment: automated detection of,
81–82; evaluation of, 103–104
September 11, 2001: collective
narrative of, 148–151
serial reproduction: Bartlett's
experiments relating to, 23,
53–58; Bartlett's theory of,
51–52; and emotions, 78–79,
80–81; myths and fairy tales as
product of, 23; and narratives'
tendency toward rationalization,
54–55; the telephone game as
example of, 22–23
shame and embarrassment: as
ambivalent emotion, 12; and
cultural misunderstandings,
135–136; laughter as response
to, 136–137; repressive function
of, 134–135; and stigmatization,
86–88, 137
Silence of the Lambs, The (film), 65
Silverstein, Shel: *The Giving Tree*,
33–34